COMPARATIVE PLANETOLOGY
WITH AN EARTH PERSPECTIVE

First International Conference on
Comparative Planetology
with an Earth Perspective

June 6–8, 1994
The Pasadena Hilton Hotel
California Institute of Technology
Pasadena, California

Sponsored by
National Aeronautics and Space Administration
Solar System Exploration Division

Hosted by
Jet Propulsion Laboratory
Office of Space Science and Instruments
California Institute of Technology

CONFERENCE CHAIRS
Dr. Moustafa T. Chahine, Jet Propulsion Laboratory
Dr. Michael F. A'Hearn, University of Maryland

SCIENCE ORGANIZING COMMITTEE
Dr. Raymond E. Arvidson, Washington University
Dr. Alexander T. Basilevsky, Vernadsky Institute
Dr. Reta Beebe, University of New Mexico
Dr. Thérèse Encrenaz, Observatoire de Paris
Dr. Christopher McKay, NASA Ames Research Center
Dr. Ronald S. Saunders, Jet Propulsion Laboratory
Dr. Heinrich Wänke, Max-Planck Institut für Chemie
Dr. Maria T. Zuber, Goddard Space Flight Center

LOCAL ORGANIZING COMMITTEE
Mr. Neil L. Nickle, Jet Propulsion Laboratory
Ms. Patricia B. McLane, Jet Propulsion Laboratory
Dr. Jeffrey B. Plescia, Jet Propulsion Laboratory
Dr. Ellen R. Stofan, Jet Propulsion Laboratory

TECHNICAL SUPPORT
Pamela Solomon, Goddard Space Flight Center

Comparative Planetology
with an
Earth Perspective

Proceedings of the First International Conference
held in Pasadena, California, June 6–8, 1994

Edited by

MOUSTAFA T. CHAHINE

Jet Propulsion Laboratory, Pasadena, CA

MICHAEL F. A'HEARN

University of Maryland, College Park, MD

and

JÜRGEN RAHE

NASA Headquarters, Washington, DC

Technical Editors

Pamela Solomon, *Goddard Space Flight Center, Greenbelt MD*
Neil L. Nickle, *Jet Propulsion Laboratory, Pasadena, CA*

Reprinted from *Earth, Moon, and Planets*
Volume 67, Nos. 1-3, 1994/1995

SPRINGER-SCIENCE+BUSINESS MEDIA, B.V.

A C.I.P. Catalogue record for this book is available from the Library of Congress.

ISBN 978-90-481-4636-9   ISBN 978-94-017-1092-3 (eBook)
DOI 10.1007/978-94-017-1092-3

Printed on acid-free paper

EARTH, MOON, AND PLANETS / *Vol. 67   Nos. 1–3   1994/1995*

COMPARATIVE PLANETOLOGY WITH AN EARTH PERSPECTIVE

Edited by M. T. CHAHINE, M. F. A.'HEARN and J. RAHE

Foreword	vii
Introduction	ix–x
WILLIAM M. KAULA / Formation of the Terrestrial Planets	1–11
RONALD GREELEY / Geology of Terrestrial Planets with Dynamic Atmospheres	13–29
S. W. BOUGHER, D. M. HUNTEN and R. G. ROBLE / CO_2 Cooling in Terrestrial Planet Thermospheres	31–33
JOHN E. BRANDENBURG / Constraints on the Martian Cratering Rate Based on the SNC Meteorites and Implications for Mars Climatic History	35–45
A. T. BASILEVSKY / Factors Controlling Volcanism and Tectonism in Solar System Solid Bodies	47–49
G. WUCHTERL / Giant Planet Formation: A Comparative View of Gas-Accretion	51–65
NOBUYA TAJIMA and YOSHITSUGU NAKAGAWA / Giant Planet Formation: Dynamical Stability of a Massive Envelope	67–69
S. K. ATREYA, S. G. EDGINGTON, D. GAUTIER and T. C. OWEN / Origin of the Major Planet Atmospheres: Clues from Trace Species	71–75
T. ENCRENAZ / The Chemical Atmospheric Composition of the Giant Planets	77–87
THOMAS R. SPILKER / NH_3, H_2S, and the Radio Brightness Temperature Spectra of the Giant Planets	89–94
A. COUSTENIS / Titan's Atmosphere and Surface: Parallels and Differences with the Primitive Earth	95–100
J. S. KARGEL / Cryovolcanism on the Icy Satellites	101–113
D. MÖHLMANN / Formation of Satellite and Ring Systems: Comparative Aspects	115–129
L. L. HOOD / Frozen Fields	131–142
E. H. LEVY / Planetary Dynamos	143–160

MICHAEL SCHULZ / Planetary Magnetospheres 161–173

L. J. ZANETTI, T. A. POTEMRA, and B. J. ANDERSON / Boundary
Determinations from Low Frequency Magnetic Field Measurements 175–178

JEFFREY N. CUZZI / Evolution of Planetary Ringmoon Systems 179–208

T. V. RUZMAIKINA / Thermal History of Planetary Materials in the Solar
Nebula 209–215

S. I. IPATOV / Migration of Bodies in the Accumulation of Planets 217–219

List of Participants 221–223

The 'Kluwer' LaTeX Style File: Instructions for Authors 225–226

Author Index 227

Volume Contents 229–230

FOREWORD

The systematic study of the planets has experienced a slow but steady progress from the efforts of a single individual (Galileo Galilei, 1564-1642) to nations that individually and collectively create whole agencies and complex infrastructures devoted to the exploration and understanding of our solar system. This quest for knowledge continues in earnest today as we attempt to understand Earth's unique place among its closest neighbors. Known diversities emphasize fractionation processes that may have occurred in the nebula during early solar system formation, and the vastly different evolutionary paths taken by the planets and their satellites. The discovery of similarities and differences among the planets has given rise to a discipline of "Comparative Planetology." Here terrestrial properties and giant planet atmospheres are viewed and probed, surface geologies are related to atmospheres and oceans, interior structures are envisioned, magnetic fields mapped, and bizarre differences in satellites and ring systems continue to enlighten, amaze and confound the detectives of planetary science.

A science organizing committee with international participation was formed to develop a conference program to address the basic issues and the fundamental processes that are common among the planets. The goals of the meeting were twofold: first the production of a reference source on comparative planetology for academia, and second, the provision of an impetus for NASA to begin a program devoted to this emerging science discipline.

The conference program accommodated seventeen invited papers and nineteen poster presentations. These proceedings represent a starting point in building reference resources for those who strive to untangle the mysteries contained in this complex corner of the universe that is home to humankind.

DR. WESLEY T. HUNTRESS, JR.
Associate Administrator for Space Science
National Aeronautics and Space Administration

Earth, Moon, and Planets **67**: vii, 1995.
© 1995 *Kluwer Academic Publishers.*

COMPARATIVE PLANETOLOGY WITH AN EARTH PERSPECTIVE

INTRODUCTION

M. T. Chahine
M. F. A'Hearn
J. H. Rahe

How the solar system was formed and how it evolved are intriguing questions basic to planetary science. Current thinking suggests that the sun and the planets were formed simultaneously through a chain of related processes. Observational data show many differences as well as similarities in the various elements of the solar system, and this implies that our planets are individual systems governed by a set of basic physical and chemical principles. However, can such principles and processes be identified and explained, and which of these processes are sufficiently general to encompass more than one planetary system? To answer these questions we must learn how physical laws work across the entire solar system and identify those common processes that led to the present state of the individual planets. This is the goal of comparative planetology – the topic of this volume.

The study of comparative planetology directly supports the goals of the NASA planetary exploration program. These goals strive to understand the origin of the solar system and the evolution of its planets, including Earth, and to describe the condition leading to the origin of life, and how life can modify planetary environments. At this time, the various planetary exploration missions have observed all the major planets of the solar system except for Pluto. The cumulative observational evidence suggests that the individual planets formed under conditions associated with the formation of the sun, which implies that they are linked by the same thread of basic physical and chemical principles.

However, the solar system has evolved considerably since its formation through numerous processes, such as the internal heating resulting from radioactive decay and bombardment by comets and asteroids. Thus, the original and the present state of the solar system are intrinsically linked. Progress toward understanding these links involves synthesis of knowledge both as a function of time (back to the origin) and space (among the planets). Today we base our knowledge on theories and hypotheses and on observations derived from over thirty years of space-based observations of the planets, and from ground-based observations extending over many centuries. Ultimately, prediction of future evolution is a definite goal. More recently a detailed in-depth examination of Earth as a planet has added to our knowledge of planetary systems, thereby making Earth a suitable reference planet for comparative planetary studies.

Earth, Moon, and Planets **67**: ix–x, 1995.
© 1995 *Kluwer Academic Publishers.*

The use of Earth data is not only convenient, it is also expedient because our current concern about the impact of human activities on our climate has profound implications to humankind. In the next few years a large amount of global data will be obtained from the International Earth Observing System (EOS) to be flown by the US, European, and Japanese space agencies to study the Earth and to predict its variability. The debate over "global change" on Earth addresses basic issues regarding the ways in which planets recycle their atmospheres and the dynamic processes which establish equilibrium between the surface of a planet and its atmosphere. Accordingly, it is likely that we will attain a better understanding of other planets such as Venus and Mars by studying them from an Earth perspective first, and then iterating the results to achieve a common consensus.

The International Conference on Comparative Planetology endeavored to follow the framework described in this introduction. The papers presented at the conference and the ensuing discussions were clearly interdisciplinary, as indicated by the group of topics covered:

1. Terrestrial atmospheres: composition and evolution
2. Giant planet atmospheres: circulation and chemical composition
3. Surface geology: physical and chemical interactions of surfaces with atmospheres and   oceans
4. Interior structure: giant planets versus terrestrial planets versus Triton and Pluto
5. Magnetic fields: dynamos versus frozen fields and magnetospheres
6. Satellites and ring systems: formation and evolution
7. Formation: giant planets versus terrestrial planets versus Triton and Pluto

The only component of the conference that is not included in these proceedings is the concluding panel discussion. Regrettably, the diversity in presentation materials did not lend itself to a formal presentation here.

FORMATION OF THE TERRESTRIAL PLANETS

WILLIAM M. KAULA
University of California, Los Angeles
Department of Earth and Space Sciences
Box 951567
Los Angeles, CA 90095, U.S.A.

Abstract

The early phases of formation in the inner solar system were dominated by collisions and short-range dynamical interactions among planetesimals. But the later phases, which account for most of the differences among planets, are unsure because the dynamics are more subtle. Jupiter's influence became more important, leading to drastic clearing out of the asteroid belt and the stunting of Mars's growth. Further in, the effect of Jupiter-- both directly and indirectly, through ejection of mass in the outer solar system-- was probably to speed up the process without greatly affecting the outcome. The great variety in bulk properties of the terrestrial bodies indicate a terminal phase of great collisions, so that the outcome is the result of small-N statistics. Mercury, 65 percent iron, appears to be a residual core from a high-velocity collision. All planets appear to require a late phase of high energy impacts to erode their atmospheres: including the Earth, to remove CO_2 so that its ocean could form by condensation of water.

Consistent with this model is that the largest collision, about 0.2 Earth masses, was into the proto-Earth, although the only property that appears to require it is the great lack of iron in the Moon. The other large differences between the Earth and Venus, angular momentum (spin plus satellite) and inert gas abundances, must arise from origin circumstances, but neither require nor forbid the giant impact. Venus's higher ratio of light to heavy inert gases argues for it receiving a large icy impactor, about 10^{-6} Earth masses from far out, requiring some improbable dynamics to get a low enough approach velocity. Core formation in both planets probably started rather early during accretion.

Some geochemical evidences argue for the Moon coming from the Earth's mantle, but are inconclusive. Large scale melting of the mantle by the giant impact would plausibly have led to stratification. But the "lock-up" at the end of turbulent mantle convection is a trade-off between rates: crystallization of constituents of small density difference versus overall freezing. Also, factors such as differences in melting temperatures and densities, melt compressibilities, and phase transitions may have had homogenizing effects in the subsequent mantle convection.

Earth, Moon, and Planets **67**: 1–11, 1995.
© 1995 *Kluwer Academic Publishers.*

1. Introduction

The terrestrial planets are distinguished from the major planets in being (1) closer to the Sun, (2) smaller, (3) made mainly of rock, and (4) separated by a marked gap, the asteroid belt. Property (1) explains properties (2) and (3): so close to the Sun that high velocities inhibited merger and fostered fragmentation, and it was hot enough that ices could not condense. But property (4) depends mainly on the innermost major planet being the largest, Jupiter, and thus having a disrupting effect by scattering smaller bodies through the next zone interior to it. The depopulation of this zone would have occurred even if a sizable planet had formed there; such a body could have been fragmented by Jupiter scattering (Wetherill, 1992). The smallness of Mars also arises from this Jupiter effect. However, between Mars and the Earth there is a qualitative difference: the "Weidenschilling Limit" (Weidenschilling, 1975). The minimum perturbation for Jupiter to eject a body from the solar system is equal to perturbation required to send a body from Jupiter to about 1.3 AU. Hence bodies going closer to the Sun than this limit are much more likely to be ejected from the solar system upon returning to the vicinity of Jupiter than are bodies not perturbed so far inward. The effect on Mercury's zone of bodies perturbed inward by the Earth and Venus is similar, in principle: there was intense fragmentation in this zone as well.

TABLE 1: Properties of the Planets

Planet	Solar Dist. AU	Mean Motion n_J	Mass/ Zone M_E/AU^2	Mass M_E	Mean Density* Mg/m^3	Metal Iron %	Atmo- sphere bars	Rota- tion Ω_E
Mercury	0.4	49.	.10	.06	5.6	65.	0.	.008
Venus	0.7	19.	.61	.85	3.95	30.	90.	.004
Earth	1.0	12.	.40	1.00	4.02	30.	1.	1.0
Moon	1.0	12.		.01	3.34	0.	0.	.037
Mars	1.5	6.3	.011	.10	3.65	15.	.01	1.0
Asteroids	2.8	2.5	.000	.00	3.4	0.	0.	—

*Reduced to a pressure of 1.0 GPa. n_J: mean motion of Jupiter; M_E: mass of Earth

The foregoing plausibilities about the main properties of the terrestrial planets in Table 1 are consistent with a scenario of hierarchical growth from an assumed initial population of small planetesimals. The earlier stages of this scenario are now well confirmed by models based on short-range dynamical interactions. However, other effects must be invoked to arrive at bodies similar to the actual planets (Lissauer,1993). This review, as appropriate to a volume on comparative planetology, will concentrate on problems associated with the later phases of the formation of the terrestrial planets, since they account for virtually all their differences in properties. This is not to say that there are no difficulties with the early phases. In particular, simple gravitational instability of dust (Safronov, 1972; Goldreich and Ward, 1973) is probably an insufficient mechanism for initial formation of planetesimals in a nebula turbulent from infalls and convection (Tscharnuter and Boss, 1993); other mechanisms of coagulation must be invoked (Cuzzi et al.., 1993; Weidenschilling and Ruzmaikina, 1994). Also, both gas drag and gravitational interaction with the nebula plausibly affected planetesimal orbits, and thence growth of embryos (Ward, 1993b). I divide the topics into (1) terminal accretion; (2) volatile gain and loss; (3) core formation; and (4) the giant impact.

2. Terminal Accretion

Models starting with a large swarm of planetesimals a few kilometers in size, such as that of Wetherill and Stewart (1993), evolve within about 10^5 years to several embryos of Moon to Mercury size in nearly circular, co-planar orbits, plus a swarm of smaller planetesimals that cannot grow because their relative velocities are enhanced to more than their escape velocities from perturbations by the embryos. This result depends on the correct equipartition of energy at encounters (Stewart and Wetherill, 1988), and has been obtained by both analytic and numerical techniques using the algorithm of Opik (1976). Fragmentation and drag are also taken into account.

To carry forward models allowing for close interactions only, larger eccentricities must be assumed. With correspondingly increased inclinations, these models take well over 10^8 years to accumulate bodies similar to the terrestrial planets (Kaula, 1990; Wetherill, 1991). Clearly, other effects need to be incorporated: at a minimum, perturbations by Jupiter. Wetherill (1992) implemented an Opik algorithm in the asteroid belt, adding the accelerations arising from sixteen resonances with Jupiter, as well as ejection by Jupiter, and shows that velocities are developed sufficient to fragment embryos of 0.02 Earth mass. However, within the Weidenschilling limit further effects must operate, whose exploration will entail considerable computational effort. Integrations of the present system obtain oscillations of the planetary eccentricities of about 0.06, almost all associated with periodicities of about 10^5 years. Hence *if* the number of terrestrial planets were doubled, there would be close encounters, leading to further enhancement of eccentricities, and thence collisions. If interactions of these embryos with a significant declining mass of nebula and planetesimals in the asteroid belt and further out are taken into account, the eventual problem may be how to limit and time this loss, in order not to enhance terrestrial planet eccentricities excessively (Ward, 1988, 1993a).

In any case, taking enhanced eccentricities as given, the consequences for the planetary embryos are severe: e.g.,with eccentricities of 0.10, approach velocities of

3 km/sec are obtained in the Earth's vicinity. Kaula (1990) started with the planets already 81 percent accumulated, plus a population of planetesimals ranging in mass up to 0.02 Earth masses. Eccentricities of both planet and planetesimal orbits were taken into account in calculating probabilities of close encounters and approach velocities in a modified Opik algorithm. The results were that proto-Mercury always received sufficient energy to cause fragmentation; proto-Mars received less, but still appreciably more than proto-Earth or proto-Venus. The range of summed colllisional energy inputs for the two larger bodies was great: from 10^6 to 2×10^7 J/kg. Angular momentum absorption was positively correlated with energy absorption, leading to spins ranging from 1 cycle/4 hrs to 1 cycle/40 hours, and obliquities up to 160°: retrograde in 30% of cases. Twelve percent of cases coupled a fast prograde rotation of one body with a slow retrograde rotation of the other.

The outcome of planetesimal interaction models such as Wetherill and Stewart (1993) appears inexorable: an inner solar system with a few dozen embryos of 0.01 to 0.06 Earth mass. Mechanisms to enhance the eccentricities of these embryos are evident, but are yet to be worked out in detail. The inevitable consequence for the terminal phases of accretion is large impacts, leading to a wide range of possible outcomes for fragmentation, loss of volatiles, spin rates, obliquities, and satellites of the proto-planets. Another important effect is the high degree of mixing among zones of the inner solar system, leading to a blurring of any compositional gradients from condensation in the nebula. The main substantive question (as distinguished from technical difficulty) is the relative phasing of terrestrial planet growth with loss of nebula, growth of Jupiter, and loss of planetesimals. At present, it appears that the losses of nebula and planetesimals must occur rather early, and perhaps the growth of Jupiter rather quickly, in order that relative velocities in the inner solar system not be enhanced too much.

3. Volatile Gain and Loss

Experimental evidences relevant to the acquisition of volatiles by the terrestrial planets include:

(1)   The H_2O content of ordinary chondrites is generally 0.2-0.3% by weight; it is almost never less than 0.1%, and can be over 10% in carbonaceous chondrites, which are probably the most abundant (Mason, 1962).

(2)   The C/H atomic ratio of chondrites varies from 0.1 to 1.0; Anders and Ebihara (1982) take 1/7 as typical

(3)   CI chondritic meteorites have not suffered as much loss of inert gases relative to cosmic abundances as the terrestrial planets (Fig. 1) (Owen et al., 1992).

(4)   Argon is trapped in amorphous ice at temperatures below 30 K, and Neon at temperatures below 20 K (Bar-Nun et al., 1988).

(5)   The ratio of H_2O solubility in magma to that of CO_2 decreases with temperature, but is always at least an order-of-magnitude greater (Mysen, 1977).

(6)   Incipient devolatilization of a CM chondrite occurred at a shock pressure of about 11 GPa, and complete devolatilization at about 30 GPa (Tyburczy et al., 1986).

Taking 3% as the H_2O of impactors (1), by the time a terrestrial protoplanet is large enough to start devolatilizing impactors (6), it will contain 1.5 Earth oceans of water. It will fully devolatilize impactors when it has six oceans (Ahrens, 1990). It has long been recognized that a terrestrial planet of Mars's mass or larger will heat up to silicate melting, if formed mainly from planetesimals of 100 km or more (Safronov, 1972). A further heating factor is that H_2O outgassed from impactors will create a steam atmosphere, which will increase until pressure sufficient to dissolve it in magma is achieved. Such an atmosphere has an infra-red opacity sufficient to maintain silicate melting temperatures at the surface, so long as appreciable infalls continue

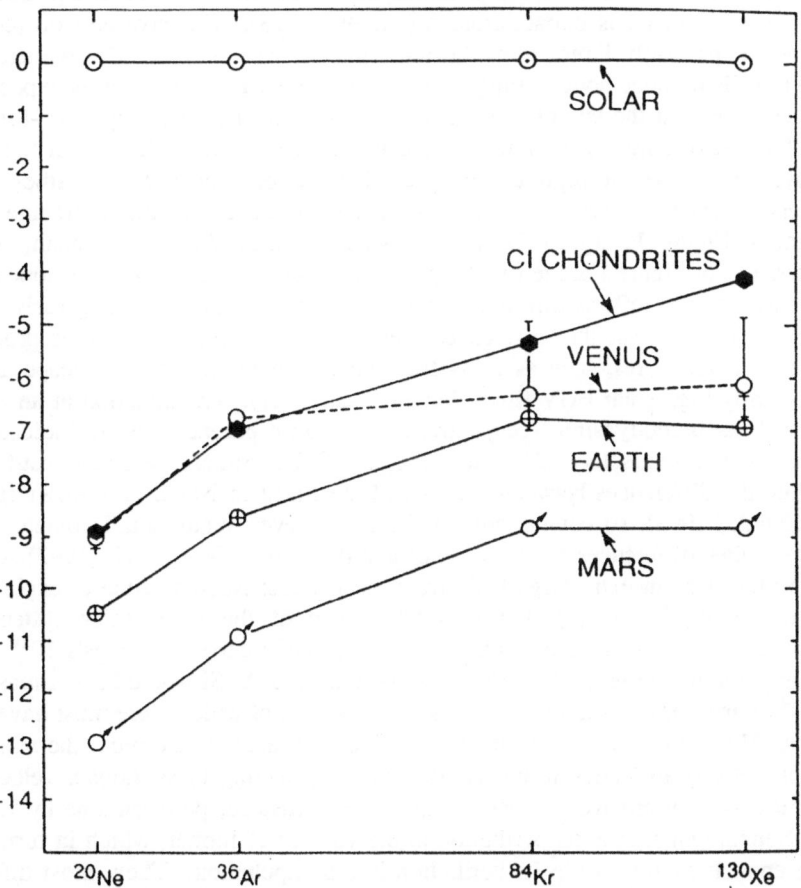

Figure 1. Inert gas abundances relative to solar abundances (normalized to silicon). The ordinate is log(abundance); the abscissa is in atomic mass units (AMU).

(Abe and Matsui, 1985). The pressure from this equilibrium atmosphere is estimated to be 30 MPa (300 bars) (Abe, 1993). If the planet is far enough from the sun and its atmosphere is dominantly H_2O, when energy inputs from infalls and the magma ocean decline the H_2O will condense and rain out. This appears to be the case for the Earth, but possibly not Venus. The problem is the carbon content of the atmosphere. If the 1/7 C/H atomic ratio applies (2), then a model which has about 3×10^{21} kg of H_2O, most of it dissolved in the magma ocean (e.g., Zahnle et al.,1988) will also have more than 10^{21} kg of CO or CO_2 , which stays in the atmosphere, since it is not soluble in magma (5). This is enough CO_2 for 15 Mpa (150 bars) pressure. While CO_2 is not as opaque as H_2O, this is more than enough to prevent condensation of water: see, e.g. figure 9 of Kasting (1988), which shows that 100 bars of CO_2 will produce a surface temperature of 480 K with solar radiation 0.8 times the present radiation at the Earth.

Hence mechanisms to remove atmospheres are needed. A compositionally insensitive mechanism is impact erosion (Ahrens, 1993). Effective erosion requires impactors appreciably larger than the atmospheric scale height. The planetesimal growth models produce bodies amply large, even though fragmentation is expected to grind them down in the late phases. The apparent loss of silicates by proto-Mercury argues for a strong impact regime. In the late phases, comet influx from the outer solar system will be at high enough velocities to be erosional for rather small impactors (Weissman, 1989). The parallelism of the Earth and Mars inert gas abundances (Figure 1) argues for a compositionally insensitive mechanism. Other characteristics of Figure 1 argue for compositionally-sensitive effects. The slope of the CI chondrite curve reflects different adsorption characteristics operating early in the formation process. But the differences therefrom of the planets requires processes peculiar to atmospheres, such as mass fractionation in hydrodynamic escape arising from the early high solar EUV flux (Hunten et al., 1987). Fractionation at an earlier stage, such as hyrdodynamic escape from a meteorite parent body, is indicated by marked trends in the isotoes of krypton (Pepin, 1991). But escape seems insufficient to explain the differences between Venus and the Earth in Ne/Kr and Ar/Kr (Pepin, 1991; Zahnle, 1993), so comets are invoked. However, many small comets would mitigate against differences between Venus and the Earth. Owen et al. (1992) suggest a single impact of an extra large body from the Uranus: Neptune zone to explain the argon excess on Venus (4), but nobody's model of the outer solar system has temperatures low enough to explain the neon. A body carrying Venus's 10^{16} kg of excess argon and having a 2.0 Mg/m^3 density and solar Ar/Si would have a mass less than 10^{-6} Earth masses-- a diameter less than 100 km, of which there must have been millions. The problem is the dynamics of delivery. It cannot be direct: the *minimum* approach velocity to Venus would be 12.9 km/sec, leading to an impact velocity of 16.4 km/sec-- very erosive, instead of accretive. Instead, perturbations by Uranus and/or Neptune must drop its perihelion to the vicinity of Jupiter, which in turn must remove enough energy to flip its perihelion into an apohelion. Then, most difficult, the Earth must quickly remove enough energy to bring its apohelion well within Jupiter's orbit. Assuming 10^8 bodies, 3×10^6 might be "dropped" to Jupiter; then 10^4

of these "flipped", and one of these "deboosted" to eventually ease into Venus. This last step seems forbiddingly improbable (Kaula, 1994).

4. Core Formation

The apparent unavoidability of magma oceans on the Earth and Venus, due to large impacts and steam atmospheres, facilitates the collection of large enough pockets of metallic iron to penetrate any solid silicate mantle below it. Thus core formation probably occurred continuously from the time planets were a few times 0.01 Earth mass, (Stevenson, 1990). But the chemical disequilibrium between the metallic core and the upper mantle suggests there was an evolution toward more oxidizing conditions as the impact rate decreased and magma oceans solidified (Newsom and Sims, 1991). These conditions apply as well to Mercury, which had the greatest energy inputs from impacts, and evidences a history of cooling since origin. Mars is more problematical; its high mantle density and high moment-of-inertia indicate more iron was retained in the mantle, while its surface topography suggests a warming history, as would occur with delayed core formation.

5. The Giant Impact

The *only* property requiring a great impact in the late phases of the formation of the Earth is the extraordinary depletion of the Moon in iron: not only must any metallic core be negligible, but also the silicate Fe/Mg ratio of the bulk Moon must be as low as that in the upper mantle, appreciably less than in chondritic meteorites. The alternative of a screening mechanism in a swarm around proto-Earth founders on the time scale of accretion of such a swarm into one body being much shorter than the time scale of infall of additional bodies from heliocentric orbits, so that most of the time the proto-lunar material is clumped in a body big enough to be insensitive to differences among infalling bodies. The high angular momentum and big satellite of the Earth can be explained by a circum-terrestrial swarm, provided that it starts early in planetary growth and the Earth's tidal dissipation factor (1/Q) is big enough (Harris and Kaula, 1975).

It should be emphasized that the largest impact into one protoplanet, the Earth, being considerably larger than the largest impact into another, Venus, is to be expected from the strongly stochastic nature of the terminal phase of accretion discussed above. Indeed, it is the Occamish hypothesis, and alternative hypotheses have the burden of explaining why it did not occur.

Characteristics of the giant impact indicated by modelling are: (1) widespread melting and volatilization, but very non-uniform; (2) the rapid merger of the core of any impactor with the core of the Earth; (3) the formation of a surrounding ring of gases and volatilized silicates; (4) the ondensation of some of these silicates in orbit,to form the proto-Moon mainly from impactor material; (5) almost complete loss of atmosphere due to heating sufficient to create a planetary wind, if not by hydrodynamic effects; and (6) the formation of a deep magma ocean. (Cameron and Benz, 1991; Melosh et al., 1993).

Quite consistent with a much greater rocky impact in Earth than in Venus formation, but not demanding it, are: the oxygen isotope identity of lunar and Earth ultramafic rocks; the low volatile content of The Moon (e.g., a K/U ratio of about 2000); the low oxidation level of the Moon (e.g., no Fe_2O_3); the evidences (KREEP, etc.) of a magma ocean on The Moon; the high angular momentum density of the Earth-Moon system; and the several volatile properties of the planets discussed above: lower inert gas retention in the Earth than in Venus; and ocean and atmosphere by outgassing from meteoritic composition, plus a minor veneer from the outer solar system.

The main objections to the giant impact, most articulately from Ringwood (1992 and earlier papers) were: consistency (within a factor of three) of lunar volcanic glasses (LVG's) with the Earth's mantle in abundances of siderophiles (including volatiles: e.g., germanium), implying thorough mixing of impactor and the Earth; (2) the improbability of such a big impact occurring so near the end of accretion that the Moon did not acquire metallic iron and siderophiles, and (3) the gross homogeneity of the Earth's mantle, which appears to forbid a deep magma ocean.

In regard to objection (1), the factor-of-three in the 3.5 Ga LVG's contrasts to the factors-of-hundreds for older basalts, and suggests that deep in the Moon are chunks that were only partly devolatilized. However, that the LVG's pyroclastic delivery indicates that they are peculiar in volatile content, and hence they may be as little representative of the bulk Moon as kimberlites are of the bulk Earth. Meanwhile, further compositional evidences of provenances of different sources than the Earth's mantle for lunar material, but a lack of volatility-associated fractionation, have been found by Humayan and Clayton (1994) and Norman et al. (1994). Partitioning in >2000 K melts, let alone >4000 K vapors, is not measurable and difficult to extrapolate. There could be more than one way to get scruffy within-a-factor-of-three similarities in siderophiles, so the lunar material need not have been at pressures comparable to deep in the Earth's mantle.

Regarding objection (2), the implausibility of the lateness is in the context of a lot of smaller bodies still impacting, but with high relative velocities, hence more likely to erode, than accrete, the Moon, which has an escape velocity of only 2.36 km/sec.

Objection (3), that a deep magma ocean would stratify, is more serious. Initially, magma ocean would be homogenous because it was highly turbulent (Tonks and Melosh, 1993), carrying crystals in suspension. The difficult question is how this system "locked up" without stratifying; how to get nonfractional crystallization, dependent on suspension in convective layers, pressure, rheology of partial melts, crystal size, and surface conditions (Solomatov and Stevenson, 1993). In the subsequent evolution of the mantle, factors such as rate-dependent phase transitions; lower melting temperatures of iron-rich silicates; and higher compressibility of melts making them denser than their solid matrix at pressures >8 GPa may have been homogenizing.

Ringwood's (1992) alternative hypothesis was the removal of proto-lunar material from the Earth by several smaller high-velocity impactors. The fundamental objection to this hypothesis is they just do not generate enough hydrodynamic effects to get the "second burn" necessary to place proto-lunar material in orbit, rather than escape or return to the Earth. Even if it were found that enough small high-velocity bodies could splash a lunar abundance of material off the Earth, there would remain the problem of explaining the differences between the Earth and Venus in volatiles and spin: many small events mitigate against differences, while one or two catastrophic large events favor them.

6. Conclusions

There appears to be a growing consensus on the scenario of the formation of the terrestrial planets, including the differences among them. However, there are many places where the processes connecting models and observations are quite unsure, and parts of the scenario are connected by loose plausibilities.

References

Abe, Y. (1993) Physical state of the very early Earth. *Lithos*, 30, 223-235.

Abe, Y. and Matsui, T. (1985) The formation of an impact-generated H_2O atmosphere and its implications for the thermal history of the Earth. *J. Geophys. Res.*, 90, C545-C559.

Ahrens, T. J. (1990) Earth accretion. In *Origin of the Earth*, H. E. Newsom and J. H. Jones, eds., Oxford, New York, 211-227.

Ahrens, T. J. (1993) Impact erosion of terrestrial planetary atmospheres. *Ann. Rev. Earth Plan. Sci. Let.*, 21, 525-555.

Anders, E. and Ebihara, M. (1982) Solar-system abundances of the elements. *Geochim. Cosmochim. Acta*, 46, 2363-2380.

Bar-Nun, A., Kleinfeld, I., and Kochavi, E. (1988) Trapping of gas mixtures by amorphous water ice. *Phys. Rev. B*, 38, 7749-7754.

Cameron, A. G. W. and Benz, W. (1991) The origin of the Moon and the single impact hypothesis IV. *Icarus*, 92, 204-216.

Cuzzi, J. N., Dobrovolskis, A. R., and Champney, J. M. (1993) Particle-gas dynamics in the midplane of a protoplanetary nebula. *Icarus*, 106, 102-134.

Goldreich, P. and Ward, W. R. (1973) The formation of planetesimals. *Astrophys. J.*, 183,1051-1061.

Harris, A. W. and Kaula, W. M. (1975) A co-accretional model of satellite formation. *Icarus*, 24, 516-524.

Humayun, M. and Clayton, R. N. (1994) The non-terrestrial origin of the Moon. *Lun. Plan. Sci. Conf.*, 25, 579-680.

Hunten, D. M., Pepin, R. O., and Walker, J. C. G. (1987) Mass fractionation in hydrodynamic escape. *Icarus*, 69, 532-549.

Jambon, A., Weber, H., and Braun, O. (1986) Solubility of He, Ne, Ar, Kr, and Xe in a basalt melt in the range 1250-1600 C, geochemical implications. *Geochim. Cosmocim. Acta*, **50**, 401-408.

Kasting, J. L. (1988) Runaway and moist greenhouse atmospheres and the evolution of Earth and Venus. *Icarus*, **74**, 472-494.

Kaula, W. M. (1990) Differences between the Earth and Venus arising from origin by large planetesimal infall. In *Origin of the Earth*, H. E. Newsom and J. H/ Jones, eds., Oxford, New York, 45-57.

Kaula, W. M. (1994) Dynamics of volatile delivery from outer to inner solar system. In *Deep Earth and Planetary Volatiles*, K. Farley, ed., American Institute of Physics, in press.

Lissauer, J. J. (1993) Planet formation. *Ann. Rev. Astron. Astrophys.*, **31**, 129-174.

Mason, B. (1962) *Meteorites*. John Wiley, New York, 274 pp.

Melosh, H. J., Vickery, A. M., and Tonks, W. B. (1993) Impacts and the early environment and evolution of the terrestrial planets. In *Protostars and Planets III*, E. H. Levy and J. I. Lunine, eds., Arizona, Tucson, 1339-1370.

Mysen, B. O. (1977) The solubility of H_2O and CO_2 under predicted magma genesis conditions and some petrological and geophysical implications. *Revs. Geophys. and Space Phys.*, **15**, 351-361.

Newsom, H. E. and Sims, K. W. W. (1991) Core formation during early accretion of the Earth. *Science*, **252**, 926-933.

Norman, M. D., Drake, M. J., and Jones, J. H. (1994) Alkali element constraints on Earth-Moon relations. *Lun. Plan. Sci. Conf.*, **25**, 1009-1010.

Opik, E.J. (1976) *Interplanetary Encounters: Close-Range Gravitational Interactions*. Elsevier, Amsterdam, 155 pp.

Owen, T., Bar-Nun, A., and Kleinfeld, I. (1992) Possible cometary origin of heavy noble gases in the atmospheres of Venus, Earth, and Mars. *Nature*, **358**, 43-46.

Pepin, R. O. (1991) On the origin and early evolution of terrestrial planet atmospheres and meteoritic volatiles. *Icarus*, **92**, 2-79.

Ringwood, A. E. (1992) Volatile and siderophile element geochemistry of the Moon: a reappraisal. *Earth Plan. Sci. Let.*, **111**, 537-555.

Safronov, V. S. (1972) *Evolution of the Protoplanetary Cloud and Formation of the Earth and Planets*. Nauka Press, Moscow, and NASA-TT-F-677, 206 pp.

Solomatov, V. S. and Stevenson, D. J. (1993) Suspension in convective layers and style of differentiation of a terrestrial magma ocean. *J. Geophys. Res.*, **98**, 5375-5390.

Stevenson, D. J. (1990) Fluid dynamics of core formation. In *Origin of the Earth*, H. E. Newsom and J. H. Jones, eds., Oxford, New York, 231-249.

Stewart, G. and Wetherill, G. W. (1988) Evolution of planetesimal velocities. *Icarus*, **74**, 542-553.

Tonks, W. B. and Melosh, H. J. (1993) Magma ocean formation due to giant impacts. *J. Geophys. Res.*, **98**, 5319-5333.

Tscharnuter, W. M. and Boss, A. P. (1993) Formation of the protosolar nebula. In *Protostars and Planets III*, E. H. Levy and J. I. Lunine, eds., Arizona, Tucson, 921-938.

Tyburczy, J. A., Frisch, B., and Ahrens, T. J. (1986) Shock-induced volatile loss from a carbonaceous chondrite: implications for planetary accretion. *Earth Plan. Sci. Let.*, **80**, 201-207.

Ward, W. R. (1988) On disk-planet interactions and orbital eccentricities. *Icarus*, **73**, 330-348.

Ward, W. R. (1993a) Disk-protoplanet interactions: torques from the coorbital zone. *Anal. NY Acad. Sci.*, **675**, 314-323.

Ward, W. R. (1993b) Density waves in the solar nebula: planetesimal velocities. *Icarus*, **106**, 274-287.

Weidenschilling, S. J. (1975) Mass loss from the region of Mars and the asteroid belt. *Icarus*, **26**, 361-366.

Weidenschilling, S. J. and Ruzmaikina, T. V. (1994) Coagulation of grains in static and collapsing protostellar clouds. *Astrophys. J.*, submitted.

Weissman, P. R. (1989) The impact history of the solar system: implications for the origin of atmospheres. In *Origin and Evolution of Planetary and Satellite Atmospheres*, S. K. Atreya, J. B. Pollack, and M. S. Matthews, eds., Univ. Arizona, Tucson, 230-267.

Wetherill, G. W. (1991) Occurrence of Earth-like bodies in planetary systems. *Science*, **253**, 535-538.

Wetherill, G. W. (1992) An alternative model for the formation of asteroids. *Icarus*, **100**, 307-325.

Wetherill, G. W. and Stewart, G. (1993) Formation of planetary embryos: effects of fragmentation, low relative velocity, and independent variation of eccentricity and inclination. *Icarus*, **106**, 190-209.

Zahnle, K. (1993) Planetary Noble Gases. In *Protostars and Planets III*, E. H. Levy and J. I. Lunine, eds., Arizona Univ., Tucson, 1305-1338.

Zahnle, K. J., Kasting, J. F., and Pollack, J. B. (1988) Evolution of a steam atmosphere during Earth's accretion. *Icarus*, **74**, 62-97.

GEOLOGY OF TERRESTRIAL PLANETS WITH DYNAMIC ATMOSPHERES

RONALD GREELEY
Department of Geology
Box 871404
Arizona State University
Tempe, AZ 85287-1404
Phone: (602) 965-7045
Fax: (602) 965-8102
E-mail: Greeley@asu.edu

ABSTRACT

Geological exploration of the solar system shows that solid-surfaced planets and satellites are subject to endogenic processes (volcanism and tectonism) and exogenic processes (impact cratering and gradation). The present appearance of planetary surfaces is the result of the complex interplay of these processes and is linked to the evolution of planets and their environments. Terrestrial planets that have dynamic atmospheres are Earth, Mars, and Venus. Atmospheric interaction with the surfaces of these planets, or *aeolian activity*, is a form of gradation. The manifestation of aeolian activity is the weathering and erosion of rocks into sediments, transportation of the weathered debris (mostly sand and dust) by the wind, and deposition of windblown material. Wind-eroded features include small-scale ventifacts (wind-sculptured rocks) and large-scale landforms such as yardangs. Wind depositional features include dunes, drifts, and mantles of windblown sediments. These and other aeolian features are observed on Earth, Mars, and Venus.

1. Introduction

Planetary surfaces are shaped or modified by various geological processes, including volcanism, tectonism, and impact cratering. Terrestrial planets that have dynamic atmospheres are further modified by agents of weathering, erosion, transportation, and deposition. For example, depending on past and present atmospheric environments, water and other fluids play an important role in surface modification of some planets. Running water is clearly important on Earth and appears to have been important in the past on Mars (Fig. 1).

In this brief review, attention is focused on processes associated with wind, or *aeolian activity*. Earth, Mars, and Venus all currently experience aeolian activity, and probably have been modified by wind throughout much of their geologic history. These planets afford the opportunity to study a basic geological process--aeolian activity--in a comparative sense, with each planet being a vast natural laboratory having strikingly different environments (Table 1). Because terrestrial aeolian processes and features

Earth, Moon, and Planets **67**: 13–29, 1995.
© 1995 *Kluwer Academic Publishers.*

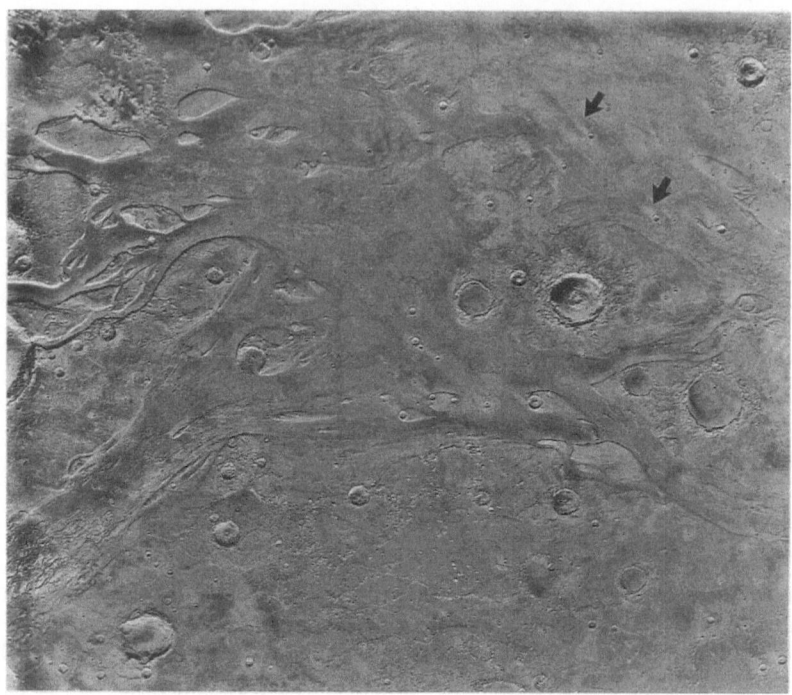

Fig. 1:
Viking Orbiter mosaic of Mars showing terrain modified by channels presumably eroded by water at a time when liquid water could exist on the surface. Also visible (arrows) are wind streaks, which are albedo patterns formed by the wind. Area shown is about 730 km by 850 km where Ares Vallis and Tiu Vallis debouch to the north (top) into Chryse Planitia. Region in the middle of the mosaic on the channel deposits has been selected as the nominal Mars Pathfinder landing site. Images in the mosaic were taken under morning illumination (sun is shining from right to left) (Viking Orbiter mosaic ASU IPF-873, centered at 20°N, 33°W).

have been studied for many years, Earth is the primary planet for comparison with Venus and Mars. However, because surface processes are much more complicated on Earth (primarily because of the presence of liquid water and vegetation) many aspects of aeolian processes that are difficult to assess on Earth are better studied on other planets.

Aeolian processes are capable of redistributing enormous quantities of sediment over planetary surfaces, resulting in the formation of landforms large enough to be seen from orbit and deposition of windblown sediments that can be hundreds of meters thick and cover thousands of square kilometers. Any process capable of effecting these

TABLE 1. Relevant properties of terrestrial planets subject to aeolian processes

	Venus	Earth	Mars
Mass (Earth = 1)	0.814	1	0.108
Density (water =1)	5.25	5.52	3.94
Surface gravity (cm s^{-2})	903	981	373
Atmosphere (main components)	CO_2	N_2, O_2	CO_2
Atmospheric pressure at surface (millibars)	9×10^4	10^3	7.5
Mean temperature at surface (°C)	480°	22°	-23°

changes is relevant to understanding the geological environment of the planets so involved. Furthermore, because aeolian processes involve the interaction of the atmosphere and lithosphere, an understanding of aeolian activity sheds light on meteorological problems.

The study of planetary aeolian processes requires a multidisciplinary, multi-task approach. Consequently, teams of geologists, engineers, and atmospheric scientists often approach aeolian research through the following tasks:

- Use spacecraft data and observations to identify the general problem and isolate specific factors for study (e.g., determine the minimum wind speeds that are required to entrain particles in different planetary environments).
- Conduct laboratory simulations for the 'Earth case' in which various parameters can be controlled (e.g., wind tunnel tests of particle threshold).
- Carry out field work to test the laboratory results under natural conditions to verify that the simulations were done correctly (e.g., observe and measure particle threshold in the field).
- Carry out laboratory experiments for the extraterrestrial cases, duplicating or simulating as nearly as possible the planetary environment involved (e.g., particle threshold tests under martian and venusian atmospheric conditions).
- Extrapolate the results to the planetary case using the laboratory results and theory (for parameters that cannot be duplicated, such as reduced gravity for Mars).

This approach not only provides a logical means for solving extraterrestrial problems, but also contributes toward a better understanding of aeolian processes on Earth.

2. Aeolian processes

Any planet that has a dynamic atmosphere with winds above a critical speed and a surface with small, loose particles has the potential for aeolian activity (Greeley and Iversen, 1985). Winds moving across a surface possess energy and can accomplish geologic work, primarily in the form of removing and transporting sand and dust. The stronger the wind, the greater the effect. The physics of windblown particles are eloquently given in the classic book by Bagnold (1941). Wind transports sediments in three modes: *suspension* (mostly silt and clay particles, i.e., smaller than about 60 μm), *saltation* (a hopping mode involving mostly sand size particles, 60 to 200 μm in diameter), and *surface creep* (particles larger than about 200 μm in diameter). Particles of these sizes can be generated on planetary surfaces from a wide variety of processes, including chemical and physical weathering, impact cratering, volcanic explosions, and tectonic deformation.

Wind threshold curves (Fig. 2) define the minimum wind speeds required to initiate movement of different sizes of particles for given planetary environments (Iversen et al., 1976a). The ability of wind to attain threshold speed is a function primarily of atmospheric density. Thus, the very low density atmosphere on Mars (the surface pressure is about 1/200 that of Earth) requires wind speeds that are about an order of magnitude stronger than on Earth. Conversely, in the very dense venusian atmosphere, very low wind speeds can set particles into motion. The density effect can be considered partly in terms of the number of gas molecules impinging on and passing over the particles to be moved; for the same amount of work to be done in the low density martian atmosphere (fewer molecules) the wind must be moving faster to achieve the same effective flux of molecules. Although this is an oversimplification, it

16

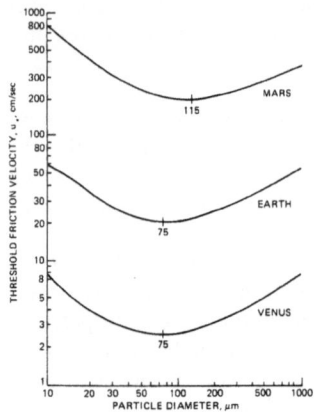

Fig. 2:
Wind threshold curves for particles of different sizes on Earth, Mars, and Venus (from Iversen et al., 1976a).

demonstrates to a first order the relationship between atmospheric density and wind velocity for particle entrainment.

As Fig. 2 shows, regardless of planet the sizes of particles most easily moved (lowest wind friction velocities) are 80 to 100 μm in diameter, or fine sand. Finer particles such as dust become increasingly difficult to set into motion under most circumstances. This results from at least two factors. First, very small grains are immersed in a laminar sublayer beneath the boundary layer and hence the wind is less effective in dislodging the grains. Second, various interparticle forces, such as cohesion and electrostatic charges (surface effects), become more pronounced with decreasing particle diameter because the surface area-to-mass ratio increases, enhancing the surface forces of attraction (Iversen et al., 1976b).

Recent laboratory experiments under martian conditions, however, demonstrate that dust settled on small (~cm) rocks can be entrained about as easily as fine sand (Greeley et al., 1994a). The rocks induce local turbulence and dust settled on the tops of rocks is placed in the turbulent boundary layer. Thus, the geologic setting and the characteristics of the surface are additional considerations of the aeolian environment.

3. Mars

Wind processes in the form of dust storms were suspected to occur on Mars even before Mariner 9 returned conclusive evidence of aeolian activity in 1971. Earthbased observations made over the last 100 years showed albedo patterns that were attributed to a variety of processes, including dust storms, as reviewed by Kahn et al. (1992), Zurek et al. (1992), and Martin and Zurek (1993). Early predictions of the wind velocities required to set particles in motion were based on knowledge of the composition and density of the martian atmosphere (Sagan and Pollack, 1969). Wind tunnel tests conducted under low atmospheric pressure in a martian simulation substantiated these estimates (Greeley et al., 1976, 1977, 1980).

Mariner 9 and the Soviet spacecraft Mars 2 and 3 arrived at Mars in 1971 during a major global dust storm, amply verifying the speculations and predictions of martian aeolian processes. After the dust cleared, the Mariner 9 cameras revealed abundant

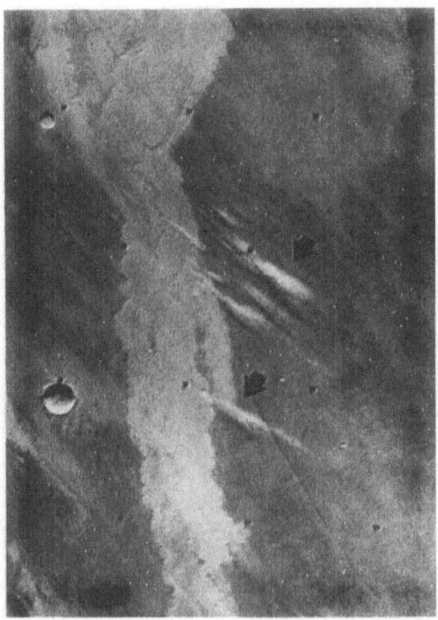

Fig. 3:
Viking Orbiter image shows small dust clouds (arrows) rising from lava flow surfaces in the southern Tharsis region of Mars (from Briggs et al., 1977); area shown is 65 km by 80 km (Viking Orbiter image 56A64).

features attributed to aeolian activity, including dunes, wind-eroded hills (yardangs), and various pits and grooves considered to be wind deflation features (McCauley, 1973). The Viking mission (1976-1981) added substantially to the catalog of martian aeolian features and provided details not previously observed (Thomas and Gierasch, 1985; Fig. 3), including the first lander images from the surface of Mars which showed fine-grained material described as aeolian drift deposits (Arvidson et al., 1978). By far, the most abundant aeolian features are wind streaks and other albedo patterns that change size, shape, and position in response to winds.

Reviews of aeolian features and processes on Mars are given by Lee et al. (1982), Thomas et al. (1981), Greeley and Iversen (1985), Greeley et al (1992), and others. Dust deposits are discussed by Christensen (1986, 1988) and others, and reviewed by Christensen and Moore (1992).

3.1. DUNES

Dunes were discovered on Mars on Mariner 9 images (Sagan et al., 1972; Cutts and Smith, 1973). Because dunes are composed of grains that saltate, and because sand is the size grain typically moved in saltation (regardless of planetary environment), it is assumed that the martian dunes (Fig. 4) are composed of grains 60 to 2000-μm in diameter (i.e., "sand").

The Viking mission provided a wealth of information on martian dunes and dune forms. One of the most striking discoveries was the huge sand sea of the north polar region. The field covers more than 7×10^5 km^2, larger than Rub Al Khali in Arabia, the

18

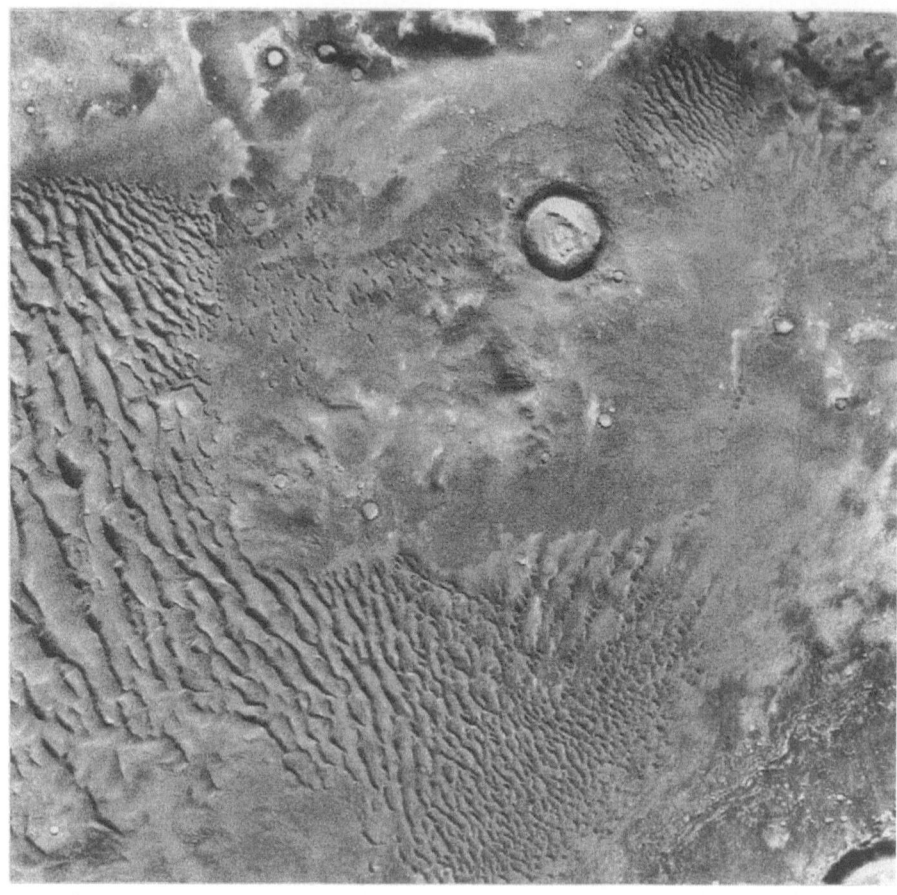

Fig. 4:
Sand dune fields on the floor of the martian crater Kaiser showing transverse dunes of a large (upper part of picture) and small (lower right) field and individual barchan dunes (above the crater). From the shape of the dunes, the prevailing wind direction at the time of dune formation was from the top of the image to the bottom; area shown is about 52 km by 52 km (Viking Orbiter image 575B60).

largest active erg on Earth. All of the martian north polar dunes are either transverse or barchan forms (Fig. 5). Mapping the dune morphologies (Tsoar et al., 1979; Breed et al., 1979) and other indicators of wind directions have enabled maps of the wind circulation pattern for the north polar area to be derived. Two major wind directions are suggested, off-pole winds that become easterly due to coriolis forces during summer,

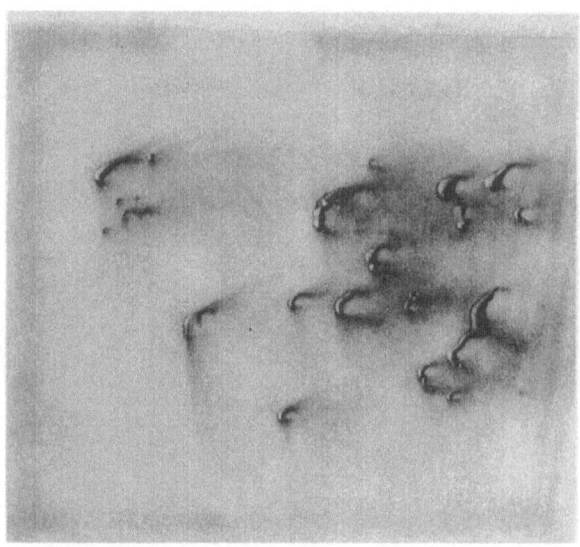

Fig. 5:

Barchan dunes of the martian north polar region; wind was blowing left to right when the dunes were active. Fine dark material emanating from the "horns" of the dunes suggested to Tsoar et al. (1979) that the dunes may be active; area shown is about 20 km by 20 km (Viking Orbiter image 544B05).

and on-pole winds that become westerly during winter. These wind patterns compare favorably with those based on models of the atmosphere (Pollack et al., 1990).

The low albedo (i.e., dark) of the north polar dunes suggests a composition other than quartz (the common sand composition on Earth), an observation fitting with the apparent lack of silicic materials on Mars (Krinsley et al., 1979). Because basaltic lavas are very common over much of Mars, including the smooth plains south of the dune field, one suggestion is that the north polar dunes are composed of windblown basaltic sand (Tsoar et al., 1979). Alternatively, studies of the colors of the dune deposits suggest that some of the material could be derived from the polar layered deposits (Thomas and Weitz 1989) or other local materials (Saunders and Blewett, 1987).

In addition to the north polar sand sea, dunes and dune fields are found in many places as isolated deposits (Fig. 4). Edgett and Christensen (1991, 1994) analyzed the thermal inertia properties of some of these deposits and estimated the particle sizes. They found an average size of 500 ± 100 μm, or medium to coarse sand, which is about twice the average size for dune deposits on Earth.

3.2. YARDANGS

Yardangs are wind-sculpted hills that have the appearance of inverted boat hulls. First discovered on Mariner 9 images (McCauley, 1973), most yardangs on Mars occur in equatorial regions, notably in the Amazonis region, Aeolis region, Ares Valles, and Iapygia. Some of the largest features are interpreted to be early-stage yardangs; they are 50 km long, 1 km wide, and 200 m high, and appear to have developed from the erosion of mesas. From studies of terrestrial yardangs, Ward (1979) concluded that the martian features are geologically young (on a martian time-scale) and probably are composed of friable rocks such as ignimibrites (many of the yardang localities are near known volcanic craters), or indurated regolith (regolith in this sense being fragmental debris

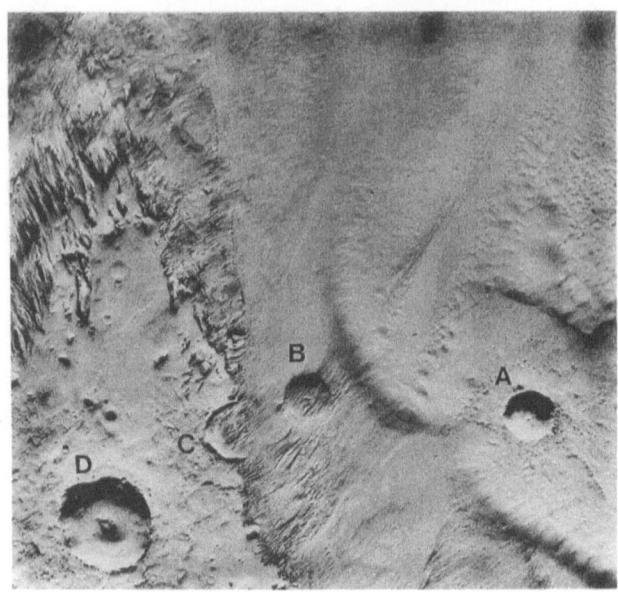

Fig. 6:
Exhumation of ancient cratered terrain (left side of image) as a younger mantling deposit (right side of image) is stripped away by the wind. Note the sequence of craters: A) fresh crater superposed on mantling deposit, B) mantled crater, C) half mantled, half exhumated crater, and D) exhumed crater. yardangs (wind-sculpted hills) form along the margins of the mantle; shown is about 75 km by 75 km (from Greeley et al., 1985) (Viking Orbiter image 438S01).

generated by impact cratering). On Earth yardangs develop by erosion of grains that are loosened by weathering processes involving liquid water; Ward suggests that on Mars (in the absence of liquid water) exfoliation, salt weathering, or freeze-thaw processes may operate, but that the net weathering rate would be slower than on Earth.

The orientations of the martian yardangs are inconsistent with wind directions predicted from global circulation model (GCM) runs simulating the atmosphere. This suggests that the yardangs formed under a different wind regime or that the yardang orientations are dominated more by structural patterns, such as joints or fractures, than by wind directions (Greeley et al., 1993).

Regardless of mode of formation or the material comprising the yardangs, the area containing the yardangs shown in Fig. 6 of the Amazonis region demonstrates the erosive potential by winds on Mars. This region originally consisted of ancient cratered terrain that was blanketed with mantling deposits of presumed windblown dust or volcanic ash. Subsequently, part of the cratered terrain has been exhumed, re-exposing the craters. The margins of the mantling blanket are being cut away by wind erosion, forming the yardangs.

3.3. VARIABLE FEATURES

Variable features were named from the Mariner 9 mission results (Sagan et al., 1972) for albedo patterns that changed their size, shape, and position with time. Crater streaks, the most common of the variable features, can occur as either light (Fig. 7) or dark forms, although 'mixed' forms are found in which both light and dark patterns occur in association with the same crater. Most investigators agree that streaks represent a

Fig. 7:
Bright wind streaks in the eastern Elysium Planitia region of Mars. Wind streaks are albedo patterns that have a contrast with the albedo of the surface on which they occur; note that the streaks in the lower right hand corner of the image are more difficult to discern because the background plains are also high albedo; area shown is about 150 km by 180 km (Viking Orbiter frame 545A54).

surface manifestation of windblown processes, such as relatively thin (~cm) deposits of particles that shift in response to winds.

Various models for streak formation have been proposed. Most models take into account the flow patterns generated by winds blowing over and around craters or other topographic features such as hills and ridges. Wind tunnel simulations and field studies show that a horseshoe vortex wraps around the crater rim and creates an erosive zone in the wake of the crater and a depositional zone in the immediate lee of the crater rim (Greeley et al., 1974). The size and shape of zones of erosion and deposition are functions of crater geometry, wind speeds, time, and other parameters (Iversen et al., 1976c). This model appears to be appropriate for explaining low albedo, or dark streaks. High albedo, or bright streaks appear to be dust deposits. Veverka et al. (1981) and Thomas et al. (1984) suggested that bright streaks form as a function of atmospheric stability, leading to dust deposition.

Crater streaks can be used to map surface wind patterns and be applied to atmospheric circulation models (Thomas and Veverka, 1979). A general circulation model (GCM) has been developed for Mars which enables near-surface wind direction and strength to be predicted as a function of season (Pollack et al., 1981, 1990). The model takes into account features such as dust opacity, surface albedo, and martian topography. Comparisons of wind streak locations and orientations visible on images provide a validation for the GCM and shed light on wind streak formation (Greeley et al., 1993). There is a good correlation between bright streaks and GCM predictions, both in terms of orientation and locations of maximum wind speeds during the southern hemisphere summer. Dark streaks, however, show poorer correspondence; this is attributed to dark streak formation related to local winds that are not well represented by the GCM.

4. Venus

Aeolian processes on Venus have been debated for more than two decades, and many investigators predicted that aeolian features would eventually be found (reviewed by Greeley and Arvidson, 1990). Although images of the surface returned from the Soviet Venera landers and measurements of near-surface winds suggested local modification of the surface by wind, definitive evidence for more widespread aeolian activity was not observed until the Magellan mission (Saunders et al., 1991).

The atmosphere of Venus is composed primarily of CO_2 with minor amounts of hydrochloric, hydrofluoric, and sulfuric acids. With a surface pressure of more than 90 bar, it has the highest atmospheric density of all the terrestrial planets (Table 1). Venus is completely enveloped in a perpetual shroud of clouds that hide the surface from viewing. Repetitive pictures of the cloud tops obtained over a period of 8 days during the flyby of Mariner 10 in 1974 showed circulation patterns and allowed wind speeds to be determined for the upper atmosphere (Murray et al., 1974). From the cloud motion patterns, it was suggested that Hadley circulation dominated the upper atmosphere of Venus. Although speeds of about 100 m s^{-1} were obtained for the upper clouds in the equatorial zone, when extrapolations were made to the surface the winds were estimated to be very sluggish.

The Soviet landers, Venera 9 and 10, measured wind speeds near the surface for two sites on Venus of 0.5 to 1 m s^{-1} at the height of the windsensors (1 to 2 m above surface). More recent measurements of windspeeds obtained by the Pioneer-Venus atmospheric probes were extrapolated to the surface and yield values of 1 to 2 m s^{-1} (Counselman et al., 1979). These values are well within the range predicted for particle threshold (Fig. 1), based on a combination of theory (Hess, 1975; Sagan, 1975) and wind tunnel experiments (Iversen et al., 1976). Venera lander images of the surface show rock fragments several cm across and larger, set in a mass of fine (<1 cm) material interpreted to be sand size or smaller (Florensky et al., 1977). This bimodal size distribution is indicative of fluid transport and because liquid water cannot exist at the extremely high temperature on Venus, it is assumed that the fluid involved is the atmosphere, or wind.

Magellan synthetic aperture radar data reveal numerous surface features that are attributed to wind processes (Arvidson et al., 1991). These include dune fields, yardangs, and the most common aeolian feature, wind streaks (review by Greeley et al., 1994c).

4.1. WIND STREAKS

Venusian wind streaks are radar backscatter patterns that contrast with the surrounding surface (Arvidson et al., 1992; Greeley et al., 1992b). Both radar-bright (high radar backscatter, generally caused by rough surfaces) and radar-dark (low radar backscatter, generally resulting from smooth surfaces, or surfaces composed of particulate material such as sand) wind streaks occur. Although streaks range from less than 5 km long to several hundred kilometers, typical streaks are about 20 km long. Streaks occur in several shapes, including plume, fan, and long-narrow forms (Fig. 8). The most abundant, informally termed "zebra" streaks, consist of multiple, alternating radar dark and bright streaks. Nearly all zebra streaks are associated with deposits inferred to be

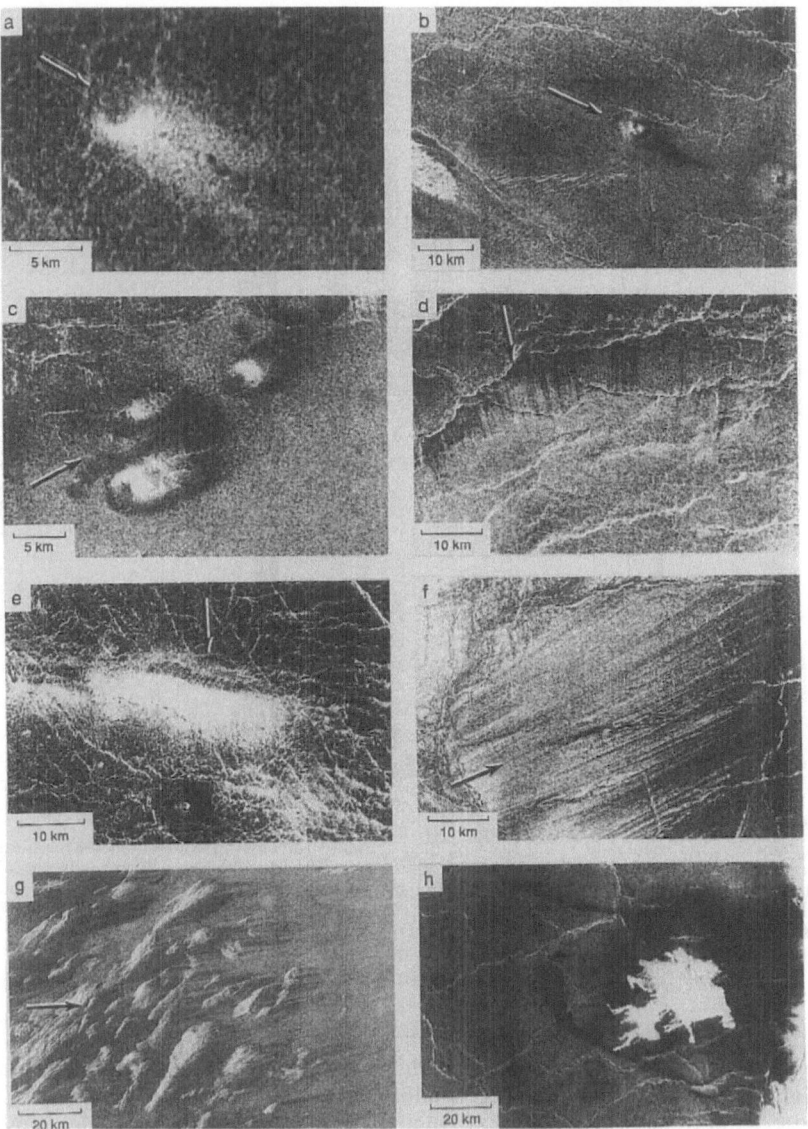

Fig. 8:

Venus wind streaks (arrows indicate downwind direction). (a) Radar-bright fan-shaped wind streak 10.5 km long associated with a small hill in eastern Niobe Planitia, centered at 36.5°N, 174.6°E (Magellan F-BIDR 1194). (b) Radar-dark fan-shaped wind streak about 10 km long associated with a small hill centered at 29.4°N, 57°E (Magellan MRPS 40983). (c) Radar-bright and -dark (mixed) fan-shaped wind streak in the Carson crater area, centered at 23°S, 344.9°E. Area shown is about 25 by 36 km (Magellan F-MIDR 23S345). (d) Linear wind streaks (radar-dark) associated with a ridge system in southern Leda Planitia, centered at 37.5°N, 65.5°E; area shown is about 44 by 64 km (Magellan MRPS 38883). (e) Transverse wind streak (radar-bright) associated with a ridge in Guinevere Planitia, centered at 26.2°N, 331.4°E; area shown is about 39 by 57 km (Magellan F-MIDR 25N333). (f) Multiple linear ("zebra") streaks in the vicinity of Mead crater, centered at 15°N, 65°E; area shown is about 44 by 64 km (Magellan MRPS 37877). (g) Multiple linear streaks (radar-dark) in western Aphrodite, centered at 0.9°S, 71.1°E; area shown is 82 by 120 km (Magellan F-MIDR 00N070). (h) Radar-dark wispy streak in eastern Sedna Planitia, centered at 37°N, 2°E; area shown is about 87 by 128 km (Magellan C1-MIDR 30N009) (from Greeley et al., 1992b).

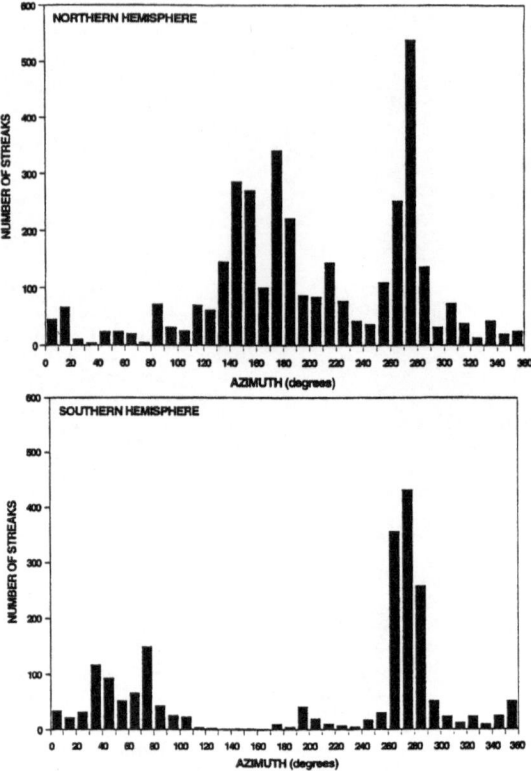

Fig. 9:
Histograms of wind streaks on Venus showing their azimuths (downwind orientation) for the northern and southern hemispheres. There is a bimodal distribution in both hemispheres. One mode is toward the equator, the other mode is toward the west. The equatorward component is consistent with Hadley circulation. The westward component is comprised of linear streaks associated with one class of venusian craters; the streaks are inferred to represent upper altitude westward zonal winds.

ejecta from young impact craters. More than 5970 wind streaks have been identified on Venus.

Similar to wind streaks on Earth and Mars (Sagan et al., 1972; Thomas et al., 1981; Greeley et al., 1989), venusian streaks are thought to be visible on radar images because of differences in the distribution of windblown particles as related to surface wind patterns. Wind tunnel experiments simulating Venus suggest that particles moved by the wind are smaller than ~1 cm, and that most would be a few hundred μm in diameter. Depending on the wind friction velocity, surfaces may be completely stripped of loose grains (leaving exposed bedrock), covered with large (>few cm) particles too massive to be removed by the wind (forming a lag deposit), or blanketed with grains transported from elsewhere and deposited in areas of low surface winds. Extremely rough surfaces, such as some lava flows, may serve as traps for wind-transported particles. Each of these surfaces would have different radar backscatter properties (Arvidson et al., 1992), depending on several considerations, including the areal extent and thickness of surficial deposits, exposed bedrock and its surface-roughness, and possible aeolian bedforms, such as dunes.

Most venusian wind streaks are associated with deposits from certain impact craters and some tectonically deformed terrains, suggesting that both of these geological

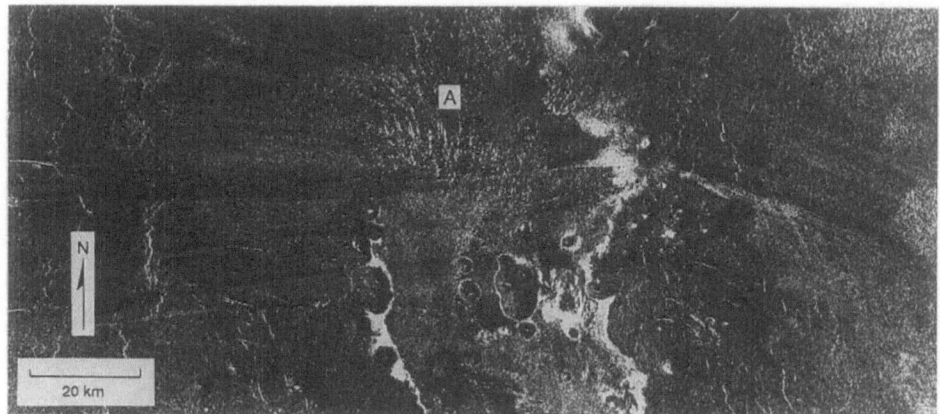

Fig. 10:
Aglaonice dune field, centered at 24.8°S; area shown is ~78 by 180 km. This dune field, indicated by the specular pattern at "A", is located within in outflow associated with the Aglaonice impact crater. Radar-dark linear streaks sweep across the area, suggesting winds from the east (right) toward the west (left). If this wind orientation is correct, the proposed dunes would be transverse forms (Magellan MRPS 34032); (from Greeley et al., 1992b).

settings provide fine particulate material that can be entrained by the low-velocity winds on Venus. Turbulence and wind patterns generated by the topographic features with which many streaks are associated can account for differences in particle distributions and in the patterns of the wind streaks. Thus, some high backscatter streaks are considered to be zones that are swept free of sedimentary particles to expose rough bedrock; other high backscatter streaks may be lag deposits of dense materials from which low-density grains have been removed (dense materials such as ilmenite or pyrite have dielectric properties that would produce high backscatter patterns).

To assess potential atmospheric circulation patterns on Venus, all wind streaks identified on Magellan images were mapped, measured, and classified (Greeley et al., 1994b). Histograms of streak orientations were plotted for the northern and southern hemispheres (Fig. 9). Orientations are given as azimuths in the downwind direction. In the northern hemisphere there is a bimodal distribution of azimuths; one mode is toward the south-southeast and the other is toward the west. The azimuths in the southern hemisphere are also bimodal with one mode toward the north-northeast and the second mode toward the west. Thus, the global wind directions inferred from the streaks are generally equatorward and toward the west.

Analysis of the orientations of venusian wind streaks suggests that the Hadley circulation proposed for the upper atmosphere extends to the surface, influencing the development of the streaks (Greeley et al., 1994b). Moreover, mapping the locations of the streaks suggests that Hadley cells extend to both polar areas of Venus.

4.2. OTHER AEOLIAN FEATURES

In addition to wind streaks, other aeolian features on Venus include yardangs (?) and dune fields. The Aglaonice dune field, centered at 25°S, 340°E, covers ~1290 km^2 and is located in an ejecta flow channel from the Aglaonice impact crater (Fig. 10). The Meshkenet dune field (Fig. 11), located at 67°N, 90°E, covers ~17,120 km^2 in a valley between Ishtar Terra and Meshkenet Tessera. Wind streaks associated with both dune fields suggest that the dunes are of transverse forms in which the dune crests are

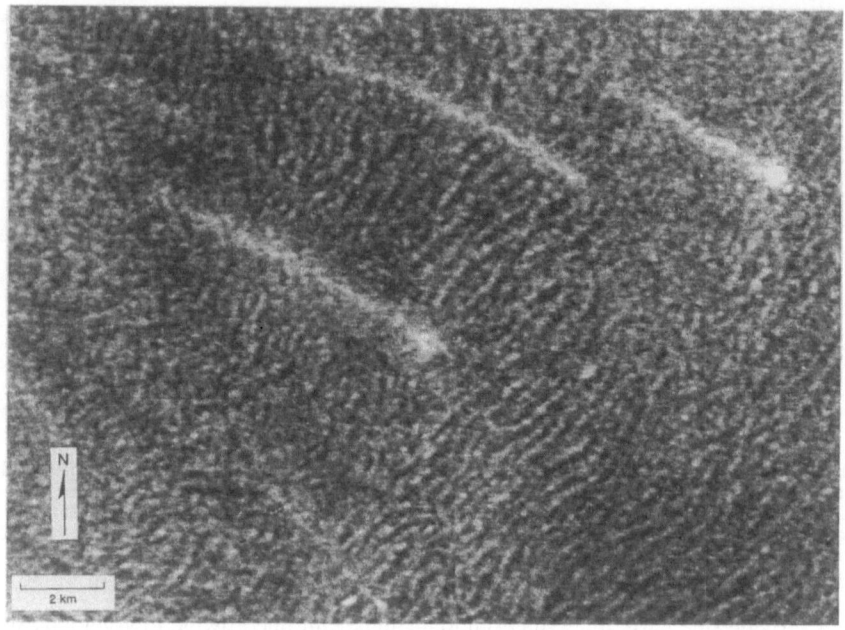

Fig. 11:
Part of the Fortuna-Meshkenet dune field (67.7°N, 90.5°E), showing transverse dunes and radar-bright
wind streaks (Magellan MRPS 39824).

perpendicular to the prevailing winds. Dunes on Venus signal the presence of sand-size
(~60 to 2,000 μm) grains.

The possible yardangs are found at 9°N, 60.5°E, about 300 km southeast of the
crater Mead. The yardang field is in a slight topographic depression between a ridged
belt to the northwest and the flank of Aphrodite Terra to the southeast. Analysis of fan-
shaped wind streaks suggests that winds are "funneled" through the depression toward
the equator. The orientations of the yardangs are consistent with this flow pattern,
suggesting that the yardangs were eroded by the same wind regime which formed the
fan-shaped streaks.

Wind streaks and dune fields were examined on both Magellan Cycle-1 and Cycle-
2 SAR images to determine if any changes had taken place in the 8-month interval
between the two sequences. No changes could be detected that could be attributed
definitively to aeolian processes. However, as noted by Weitz et al. (1994), bright and
dark patches in several regions of the southern hemisphere were found to be visible on
SAR images of one cycle, but not the other. These patches are suggested to be zones of
microdunes in which the steeper, slip face produced a higher backscatter return when
viewed from one direction (i.e., one cycle) but when viewed from the opposite direction
(i.e., opposing cycle), the gentle, windward face produced little backscatter.

Microdunes were first proposed on Venus from analysis of Venera images
(Florensky et al., 1977). Wind tunnel simulations under venusian conditions generated
microdunes of 10-15 cm wavelength and a few cm high which had all the characteristics
of full-scale dunes, including internal bedding structures. Weitz et al. (1994) suggested
that the size and geometry are appropriate for the detection of microdunes on Magellan
images as a function of "look" direction.

5. Summary and conclusions

Terrestrial planets that have dynamic atmospheres are Earth, Mars, and Venus. Depending on environments, these planets experience a variety of surface-modifying processes including weathering, erosion, and deposition. Winds currently shape all three planets and geological evidence shows that aeolian processes have operated throughout their "visible" history.

Wind erosional landforms (yardangs, eroded rocks, deflation pits), depositional landforms (dunes, mantling deposits), and wind streaks are seen on Earth, Mars, and Venus. Determining the location and age of these features gives insight into geologic histories and the interactions of the atmosphere and the surface.

Some of the key problems to be addressed for Mars and Venus in the future include:
- Determining the particle size and geographic distribution of windblown materials,
- Determining if and how cycles of windblown materials operate,
- Determining if sand dunes are currently active,
- Determining rates of erosion and deposition,
- Determining the composition(s) of windblown material,
- Determining the time scales of wind streak evolution.

Some of these issues will be addressed for Mars on forthcoming missions such as Mars Global Surveyor, Mars Pathfinder, and Mars-96. With no currently approved missions for Venus, many of these issues will not be addressed for years; however, the Magellan data set undoubtedly has much to yield with further analysis.

References

Arvidson, R.E., Binder, A.B., and Jones, K.L. (1978) The surface of Mars, *Sci. Amer.* **128**, No. 3, 76-89.
Arvidson, R.E., Baker, V.R., Elachi, C., Saunders, R.S., and Wood, J.A. (1991) Magellan analysis of Venus surface modification, *Science* **252**, 270-275.
Arvidson, R.E., Greeley, R., Malin, M., Saunders, R.S., Izenberg, N., Plaut, J.J. and Stofan, E. (1992) Surface modification of venus as inferred from Magellan observations of plains and tesserae, *J. Geophys. Res.* **97**, 13,303-13,317.
Bagnold, R.A. (1941) *The Physics of Blown Sand and Desert Dunes*, Methuen and Co., London, 265 p.
Breed, C.S., Grolier, M.J., and McCauley, J.F. (1979) Morphology and distribution on common 'sand' dunes on Mars: Comparison with Earth. *J. Geophys. Res.* **84**, 8183-8204.
Briggs, G., Klaasen, K. Thorpe, T., and Wellman, J. (1977) Martian dynamical phenomena during June-November 1976: Viking Orbiter imaging results. *J. Geophys. Res.* **82**, 4121-4149.
Christensen, P.R. (1986) Regional dust deposits on Mars: Physical properties, age, and history, *J. Geophys. Res.* **91**, 3533-3545.
Christensen, P.R. (1988) Global albedo variations on Mars: Implictions for active aeolian transport, deposition, and erosion, *J. Geophys. Res.* **93**, 7611-7624.
Christensen, P.R. and Moore, H.J. (1992) The Martian surface layer, in H.H. Kieffer, B.M. Jakosky, C.W. Snyder, and M.S. Matthews (eds.), *Mars*, Univ. of Arizona Press, Tucson, pp. 686-729.
Counselman, C.C., Gourevitch, S.A., King, R.W., Loriot, G.B., and Prinn, R.G. (1979) Venus winds and zonal and retrograde below the clouds, *Science* **205**, 85-87.
Cutts, J.A. and Smith, R.S.U (1973) Eolian deposits and dunes on Mars, *J. Geophys. Res.* **78**, 4139-4154.
Edgett, K.S. and Christensen, P.R. (1991) The particle size of Martian aerolian dunes, *J. Geophys. Res.* **96**, 22,765-22,776.
Edgett, K.S. and Christensen, P.R. (1994) Mars aeolian sand: Regional variations among dark-hued crater floor features, *J. Geophys. Res.* **99**, 1997-2018.
Florensky, C.P., Ronca, L.B., Basilevsky, A.T., Burba, G.A., Nikolaeva, O.V., Pronin, A.A., Trakhtman, A.M., Volkov, V.P., and Zazetsky, V.V. (1977) The surface of Venus as revealed by Soviet Venera 9 and 10, *Geol. Soc. of Amer. Bull.* **88**, 1537-1545.
Greeley, R. and Arvidson, R.E. (1990) Aeolian processes on Venus. *Earth, Moon, and Planets* 50/51, pp. 127-157.
Greeley, R. and Iversen, J.D. (1985) *Wind as a Geological Process*, University of Cambridge Press, Cambridge, 333 p.
Greeley, R., Iversen, J.D., Pollack, J.B., Udovich, N., and White, B. (1974) Wind tunnel simulations of light and dark streaks on Mars, *Science* **183**, 847-849.

CO₂ COOLING IN TERRESTRIAL PLANET THERMOSPHERES

S.W. BOUGHER AND D.M. HUNTEN
Lunar and Planetary Laboratory
University of Arizona
Tucson, AZ 85721

R.G. ROBLE
NCAR/High Altitude Observatory
Boulder, CO 80307

The comparative approach to planetary problems is becoming increasingly fruitful as new information from various planetary atmospheres is assimilated. As an example, it is clear that the important problem of CO_2 cooling in the Earth's lower thermosphere is closely tied to the thermospheric heat budgets of Venus and Mars.

The thermospheres of Earth, Venus, and Mars are cooled by a combination of molecular/eddy conduction and infrared (IR) cooling processes. Molecular conduction cools the upper thermosphere by transferring the EUV or auroral heat down-gradient toward the mesopause, where it is effectively radiated to space by an IR active cooling agent. The height and temperature of the mesopause are controlled by the effectiveness of the IR cooling processes. For all three planets, this IR cooling results primarily from CO_2 emission at 15-μm. Atomic oxygen collisions are known to be especially effective in exciting CO_2 vibrational states, resulting in enhanced CO_2 15-μm emissions and cooling at thermospheric heights where non-Local Thermodynamic Equilibrium (NLTE) prevails. The effectiveness of this enhancement process depends upon the relative O densities at a given pressure level plus the collisional energy transfer rate coefficient (CO_2-O) specified. The importance of this mechanism for enhancing CO_2 cooling on Earth and Venus has been debated since 1970. For Venus, a longstanding dayside heat budget problem can only be resolved by understanding the relative role of CO_2 cooling [1,2,3]; the Venus atomic oxygen density is well measured over the solar cycle. For the Earth, the lower thermosphere heat budget is difficult to understand without a thorough characterization of the variability of atomic oxygen and CO_2 cooling in that region.

Progress has been made recently that improves our understanding of CO_2 cooling processes in the Venus, Earth, and Mars lower thermospheres. First, the corresponding CO_2-O relaxation rate has been measured in the laboratory at room temperature for the first time (k = 1-2 x 10^{-12} cm³/sec). Also, derived values of this relaxation rate are becoming better constrained (1-6 x 10^{-12}) and are based primarily upon analyses of Earth CO_2 radiance and absorption measurements (see Table 1 and references therein). We see that these recently measured and derived values are not entirely consistent. Furthermore, the temperature dependence of this rate over 200-400 K has yet to be measured. This rate must be consistently incorporated into calculations of all three terrestrial planet thermospheres.

Earth, Moon, and Planets **67**: 31–33, 1995.
© 1995 *Kluwer Academic Publishers.*

CONSTRAINTS ON THE MARTIAN CRATERING RATE BASED ON THE SNC METEORITES AND IMPLICATIONS FOR MARS CLIMATIC HISTORY

JOHN E. BRANDENBURG
Research Support Instruments
635 Slaters Lane Suite G101
Alexandria Va. 22314

Abstract

Two constraints placed upon the cratering flux at Mars by the SNC meteorites are examined: crystallization ages as a constraint on surface ages and cosmic ray exposure ages and number of impacts as a constraint on absolute rates. The crystallization ages of the SNC meteorites appear to constrain the Martian cratering rate to be 4xLunar or more if the parent lavas are in the north of Mars and the number of SNC ejecting impacts are small. If the SNCs result from a single impact that formed the Lyot basin then the cratering rate must be at least 7xLunar or higher to produce a basin age less than the SNC crystallization age because the basin ages are themselves determined by crater counting. Assuming multiple uncorrelated impacts for SNC ejection from Mars over 10 million years a cratering rate of approximately 4xLunar is also found for ejecting impacts that form craters over 12km in diameter. Therefore, both crystallization ages and ejection ages and number of impacts appear consistent with a 4xLunar cratering rate at Mars. The effect on Martian chronologies of such a high cratering rate is to place the SNC crystallization ages partly within the epoch of channel formation on Mars and to extend this liquid water epoch over much of Mars history.

1. Introduction: The SNCs and Martian Chronologies

The SNC (Shergotty-Nakhala-Chassigny) meteorites are believed to be samples of Mars surface lavas that are secondary fragments from a small number of impacts on the Martian surface. The number of impacts ranges from three impacts (Bannin *et al.*, 1992) based on an apparent clustering of SNC compositions and cosmic ray exposure ages into three groups, to one large impact (Vickery and Melosh, 1987), believed to be the impact that formed the basin Lyot, near Deuteronilus Mensa. However, the measured crystallization ages of the SNCs range from .2-1.3 Gyr (Billion years) which creates a discrepancy with conventional cratering chronologies (Neukum and Hiller, 1981, Neukum and Wise, 1976, Hartmann *et al.*, 1981) used to date the surface lava units and large impact basins on Mars.

Earth, Moon, and Planets 67: 35–45, 1995.
© 1995 *Kluwer Academic Publishers.*

TABLE 1. The ages of various volcanic regions and features under various assumed values of the CCF.

Geologic Province	Crater Density Relative to Avg Lunar Maria	Estimated Crater Retention Age in billions of years (b.y.)		
		4xLunar	2xLunar	1xLunar
Central Tharsis volcanic plains	0.1	0.06	0.3	1.0
Olympus Mons	0.15	0.1	0.4	1.1
Extended Tharsis volcanic plains	0.49	0.5	1.6	3.3
Elysium volcanics	0.68	0.7	2.6	3.5
Isidis Planitia	0.76	0.8	2.8	3.6
Solis Planum volcanic	0.9	0.9	3.0	3.7
Chryse Planitia volcanic plain	1.1	1.2	3.2	3.8
Lunae Planum	1.2	1.3	3.2	3.8
Noachis ridged plains	1.3	1.7	3.3	3.8
Tyrrhenum Patera volcano	1.4	1.8	3.4	3.8
Tempe Fossae faulted plains	1.6	2.3	3.4	3.8
Volcanic plains on Hellas south rim	1.7	2.6	3.5	3.8
Alba shield volcano	1.8	2.6	3.5	3.8
Hellas floor	1.8	2.6	3.5	3.8
Syrtis Major volcanic plains	2.0	2.0	3.6	3.9
Heavily cratered plains				
− small D (<4 km)	1.4	1.8	3.4	3.8
− large D (>64 km)	13.0	3.8	4.0	4.2

Source: Hartmann et al. 1981.

SNCs are much larger under an assumed 4xLunar CCF at Mars than under 2xLunar is seen easily by comparing the areas of volcanic stratigraphies falling within the SNC crystallization ages. Volcanic terrains cover approximately 60% of Mars surface. Under the Hartmann-Tanaka chronology, which assumes a 2xLunar CCF, volcanic stratigraphies ranging approximately from the Upper Amazonian to the Early Amazonian 0-1.8 Gyr are available to serve as source regions for the .2-1.2 Gyr SNCs. These stratigraphical regions occupy 16% of Mars surface or 24% of the total volcanic terrain. If, however, the CCF is 4xLunar then the stratigraphies from the Upper Amazonian through the Upper Hesperian fall into the approximate age range of the SNCs and these stratigraphies occupy approximately 24% of Mars surface or 40% of the total lava terrain. The probability of three random impacts into such a larger area is 3 times greater under 4xLunar than into the smaller area under 2xLunar. Thus if the CCF is allowed to rise to 4xLunar, the upper limit of the range considered reasonable by Hartmann et al. (1981), then almost half of Mars lava surfaces can serve as the source region for the SNCs and the discrepancy between surface ages and the SNCs essentially vanishes. Therefore, the SNCs constrain the CCF within the limits of 1xLunar to 4xLunar to be most probably the upper limit of 4xLunar, since this gives the largest possible source area for the SNCs.

However, this explanation requires that a sample of other regions of Mars surface be found, particularly of the very old southern highland lavas, representing the impacts that occur in that region even at lower efficiency. A meteorite of Martian origin, ALH84001, apparently

A 2x LUNAR CCF GEOCHRONOLOGY OF MARS

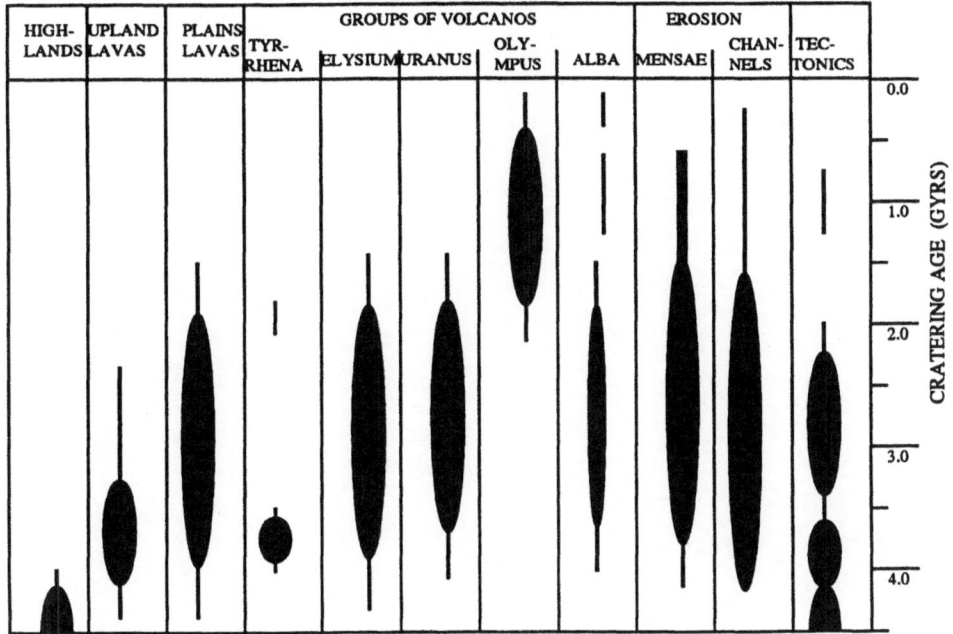

Fig. 2:
A 2xLunar CCF chronology, adapted from Neukum and Hiller Model -II, (Neukum and Hiller 1981). Note that this Lunar CCF creates a less "Lunar" geochronology for Mars , with widespread geologic activity occurring until the middle of Mars geologic history.

from the Southern Highlands, and being quite distinct mineralogically from the SNCs, has recently been identified (Middlefehlt,1994). It is apparently quite ancient, ~ 4.5Gyr crystallization age (Jagoutz E.,1994), and has an older ejection age of 15 Myr (Eugster, 1994). With the discovery of this new Martian meteorite the dichotomy of surface ages on Mars, long noted from crater statistics, is now being reproduced as a dichotomy of crystallization ages in the Martian meteorites.

It should be noted that an alternative explanation for the surface age-SNC age discrepancy, under the assumption of multiple SNC source impacts, is that surface ages are somewhat decoupled from ages of actual rocks at the surface. This is the TWY lava hypothesis, which proposes that widespread thin young lavas form the actual surface layer on Mars and do so in a manner that does not disturb crater statistics. While it has always been reasonable to consider that crater statistics can only give a gross structural age of a terrain on a planetary surface, and that small overlying deposits may occur that are much younger, the TWY lava hypothesis proposes that this phenomenon is both global and subtle on Mars. The problem with this hypothesis is that it requires widespread volcanic activity which is both recent and finely tuned. Lava flows must be numerous but small and not flood any large area to any depth. These would appear to be contradictory requirements and propose phenomenon that have not been seen either on the Moon or Earth, two bodies where considerable "ground

TABLE 2. Impacts on SNC-Age Lavas for CCFs Derived from
Hartmann-Tanaka in 10Myr

Crater Dia.	1xLunar	2xLunar	4xLunar
8km	.67	3.4	8
12km	.3	1.4	4
16km	.2	.8	2

TABLE 3. Impacts on Older Lavas for CCFs Derived from Hartmann-
Tanaka in 10Myr

Crater Dia.	1xLunar	2xLunar	4xLunar
8km	5	8	13
12km	2	4	6
16km	1	2	3

three impacts occurred within 10Myr on these younger lava units of Mars and that the
ejecting impacts produced craters of 12km diameter or greater, as was estimated to be capable
of ejecting .5 meter rocks into space by Vickery and Melosh (1987).

As can be seen in Tables 2 and 3, the rate of 4xLunar gives reasonable agreement for a 12
km diameter minimum size ejection craters, since, under this CCF, approximately 18 impacts
could be expected on the whole of surface of Mars in 10Myr with 4 of those impacts on
young Martian lava terrains of the right age range. Estimated rates are also shown for craters
of 8 km and 16 km to show the sensitivity to assumed ejection crater size. The 4 impacts into
young lavas at 12km or greater diameter also requires approximately 14 impacts on other
terrains for the same period, so that a lower efficiency of ejection must be assumed for
impacts in such terrains. Since many of these terrains are fluvial and eolian deposits or even
ices, low efficiency of ejection of these eroded materials from Mars is not surprising. The
older highland lavas probably resemble highland breccias from the Moon and as does the
newly discovered ALH84001 and likewise may not be ejected as efficiently as fresh lavas.
The lavas older than the SNCs cover 34% of Mars and they would absorb 6 impacts within
the same 10Myr under 4xLunar and would be expected to contribute some ejections even at
lower efficiency. So these missing Martian meteorites create a problem for the 4xLunar
model. The recently discovered Martian origin of ALH84001 (Middlefehlt, 1994) may
provide us with an example of such an impact, though its ejection age is slightly outside the
10Myr interval of interest.

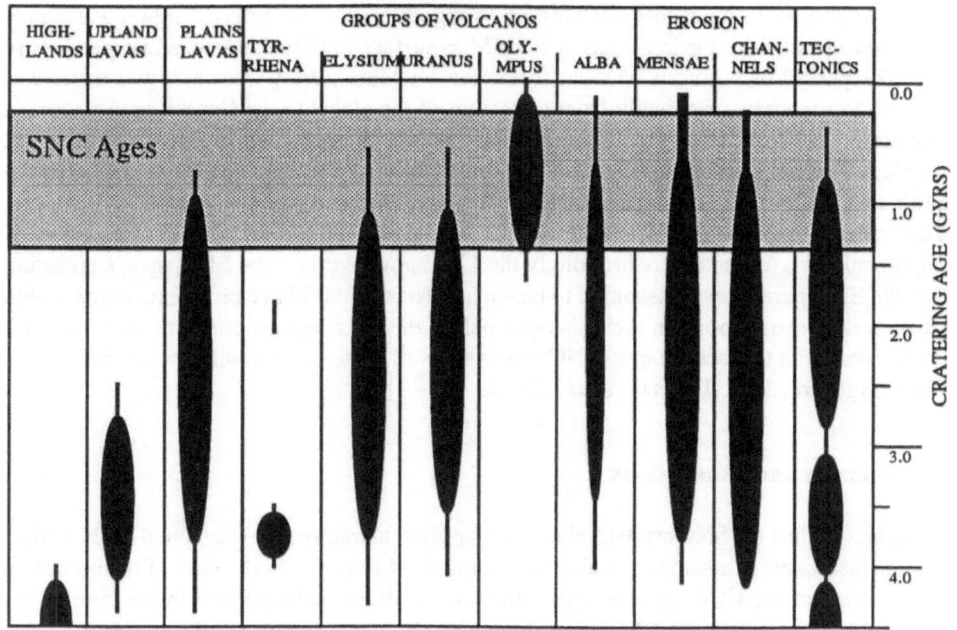

Fig. 3:
   A 4xLunar CCF chronology, after Neukum and Hiller Model -II, (Neukum and Hiller 1981). Note
   that this 4xLunar CCF creates a Terrestrial" geochronology for Mars, with widespread geologic activity
   occurring throughout Mars geologic history. In particular the histograms showing erosion, indicative of the
   presence of liquid water on Mars surface, last until late in Mars history and intrude on the epoch of the SNC
   ages.

It should be noted that 2xLunar Hartmann and Tanaka chronology gives 9 impacts into the surface of Mars at 12km or greater diameter with only 1.5 impacts into lavas in the SNC age range. The 2xLunar also predicts 7.5 impacts into other terrains of which 4 are impacts into ancient lavas. Given the low numbers of impacts and thus poor statistics the probability favors the 4xLunar model, but only marginally, over the 2xLunar model.

Thus the estimated CCF from this method would be approximately 4xLunar subject to the conditions discussed, with the fact that statistics are poor being kept very much in mind. It must be emphasized that this calculation has several other sources of uncertainty: the rate estimate is sensitive to the estimated minimum diameter of the crater left by the ejecting impact and this minimum diameter is poorly constrained, the estimated number of ejecting impacts is not well known, and the estimate of surface extent of candidate SNC source terrains in the proper age range is subject to error, given that the age of the terrains must vary with the assumed CCF. Despite these problems however, the CCF estimated by this method is consistent with the CCF found from considering only crystallization ages.

5. Implications for Mars Climatic History

If the CCF at Mars is 4xLunar or higher, and this will dramatically affect geochronologies and models based on them. An immediate scientific result of a CCF at Mars that is 4xLunar or greater, is that Mars becomes a much more geologically active and dynamic planet throughout most of its history. In addition to vigorous volcanic activity, liquid water may have existed and moved in large quantities on Mars surface until late in its history (see Figure 3). The period of vigorous channel formation on Mars , the LWE or liquid water era, appears to run stratographically from the origin of the planet to the Hesperian-Amazonian transition. It is at the end of the Hesperian age that the massive floods near the terminus of the Vallis Marineris occurred, which dwarf similar floods known to have occurred on Earth. With a CCF of 4xLunar the Hesperian era and the LWE end at approximately .9 Gyr ago. This means that the LWE on Mars may have lasted 3.5Gyr or most of Mars history. Importantly, in a 4xLunar geochronology the LWE now overlaps the SNC ages, suggesting that the SNC parent lavas, assumed to be surface rocks, could have come into contact with liquid water at some point in their histories before the ejecting impact occurred. This result is consistent with the fact that most SNCs show signs of preterrestrial liquid water alternation (Gooding *et al.*, 1991,Treiman *et al.*, 1993).

6. Summary and Conclusions

It can be seen that the SNC crystallization ages apply constraints to the assumed CCF at Mars that are insensitive to assumptions about the number of impacts. In the case of the minimum of one impact the CCF appears constrained to be above 7xLunar and in the case of the maximum of 3 impacts the CCF appears constrained to 4xLunar by surface ages. One also arrives at approximately 4xLunar by direct measurement of CCF for ejection produced craters of 12km diameter or greater, although this latter calculation has many sources of possible error. This result for the CCF pushes the epoch of abundant liquid water on Mars surface into the band of SNC ages and is consistent with signs of preterrestrial water alteration found on many SNCs. The source of the greater than Lunar CCF at Mars is readily explainable because of the proximity of Mars to the asteroid belt, now believed to be the source of most Mars impactors. Recent assays of Mars crossing asteroids performed by Gene Shoemaker and colleagues are consistent with the 4xLunar CCF at Mars (Shoemaker,1994). Mars geochronologies, all of which are dependent on the CCF value at Mars, would be profoundly altered if the 4xLunar result is confirmed and with it our concept of Mars. In particular the Lunar Mars concept, dating from the early Mariner probes, may pass away, to be replaced by a New Mars, a dynamic active planet with liquid water erosion and widespread volcanism for most of its history.

7. Acknowledgment

The author acknowledges many helpful conversations with Allan Treiman concerning SNCs, and the suggestion of the tables comparing various chronologies.

8. References

Bannin A., Clarke B.C., and Wanke H.(1992),Surface chemistry and minerology, in HughH. Kieffer, Bruce M. Jakosky, Conway W. Snyder, and Mildred S. Matthews (eds.), *Mars*, TheUniversity of Arizona Press, Tucson & London (1992) pp. 610-611.

Bogard, D. D., Nyquist, L. E.,and Johnson, P. (1984) Nobel Gas contents of shergottites and implications for the Martian origin of SNC meteorites, *Geochim. Cosmochim. Acta.*,48,1723-1740.

Eugster O. (1994) Orthopyroxenite ALH 84001: ejection from Mars 15Ma,*Meteoritics*, 29, 464(abstract).

Gooding,J.L.,Wentworth,S.J. and Zolensky,M.E., (1991), Aqueous alteration of the Nakhala meteorite, *Meteoritics*, 26, 135-143.

Hartmann W. K.,Strom R.G.,Weidenschilling S.J., Blasius K.R., Woronow A.,Dence M.R.,Grieve,J. Diaz,Chapman, E.M. Shoemaker, and Jones K.L. (1981) Chronology of planetaryvolcanism by comparitive studies of planetary cratering. in Balsaltic Volcanism Study Project(eds.) Chapter 8 *Basaltic Volcanism on the Terrestrial Planets* Pergamon Press, NewYork,pp.1049-1128.Pergamon) pp.1049-1127.

Jagoutz E.(1994) A red carbonate bearing SNC meteorite-radiometric ages on ALH84001, *Eos*, 75, 398(abstract)

Middelfehldt, D.W.(1994) ALH84001,a culminate orthopryoxenite member of the SNC meteorite group. *Meteoritics*, 29, 214-221.

Mouginis-Mark, P.J. , McCoy T. J., Taylor G.L., Kiel, K.(1992) Martian parent craters for the SNC meteorites. *Jour. Geophys.Res.*, 97, E6, p10213-10225.

Neukum, G. and Hiller, K. (1981) Martian Ages. *Jour. Geophys. Res.* 86,3097-3127.
Neukum, G. and Wise, D. U.(1976) Mars: a standard crater curve and possible new time scale, *Science*, 194, 1381-1387.

Plescia, J.B.,(1990) Young flood lavas in the Elysium region, Mars. *Lunar Planet. Sci.* XXI:969-970(abstract)

Vickery, A.M., and Melosh, H.J. (1987), The large crater origin of SNC meteorites,*Science*, 237, 738-743.

Shoemaker E.M. (1994) Private Communication.

Tanaka K.L. (1986) "The Stratigraphy of Mars" Proc. Lunar Planet. Sci. Conf. 17, *Jour.Geophys. Res.* Suppl. 91:E139-E158.

Treiman A. H.,Gooding J.L., and R.A. Barrett, (1993) Preterrestrial aqueous alteration ofthe Lafayette (SNC) meteorite, *Meteoritics*, 28, 86-97.

FACTORS CONTROLLING VOLCANISM AND TECTONISM IN SOLAR SYSTEM SOLID BODIES

A. T. BASILEVSKY

A Panel Discussion

Following the guidelines given by the leader of the panel discussion, Dr. Moustafa Chahine, your attention is directed to the attempt made by myself and Mikhail A. Kreslavsky, Kharkov Astronomical Observatory, Kharkov, Ukraine, to look for fundamental laws which control the formation and evolution of the solar system solid bodies. This work was published in a Russian journal [1], and thus is almost unknown abroad, although it has been translated into English [2]. We have attempted to find factors which control volcanism and tectonism on different solid bodies: terrestrial planets and their satellites, asteroids, and satellites of the giant planets. The terms "volcanism" and "tectonism" are used here in a broad sense. Volcanism includes eruptions of silicate melts on the terrestrial planets, sulphur on Io, and water-rich liquids on icy satellites of the giant planets. Tectonism is defined as the deformation of a body's outer layers, excluding those produced by meteoroid impacts. We used television images and radar images to classify the bodies into several categories depending on whether they have (1) only heavily cratered ancient surfaces, or whether the heavily cratered terrain on them was (2) slightly, (3) significantly, or (4) completely resurfaced by volcanism and tectonism. At the time of our study we had images of 28 solid bodies. Recent images of asteroids Gaspra and Ida obtained by the Galileo Spacecraft now extends that number to 30.

At the time of our analysis it was quite evident that the degree of post-heavy-bombardment (i.e., late) volcanic and tectonic resurfacing correlated with planet size. It is reasonable because the energy source both for volcanism and tectonism is the interior heat resulting from K, Th, and U radioactive decay, and from gravitational energy. In both cases the heat generation is proportional to the mass of the body (R^3, where R is the body's radius) while heat is being lost through the body's surface area (R^2). This results in proportionality of endogenic heat to the bodies' radii. It was also known that some satellites of the giant planets, despite their relatively small sizes, display volcanic and tectonic resurfacing which has evidently been driven by tidal heating. Tidal heating and several other factors depend on the radius ($\sim R^5$) and its orbital period (T^{-6}). This is why we created symbols showing degrees of late volcanic and tectonic activity on the accompanying diagram whose logarithmic axes are radius (R) and orbital period (T).

You will note on the diagram that symbols of the bodies with "significant," "moderate-to-weak," and "no" late endogenic activity are clustered into three fields.

Earth, Moon, and Planets **67**: 47–49, 1995.

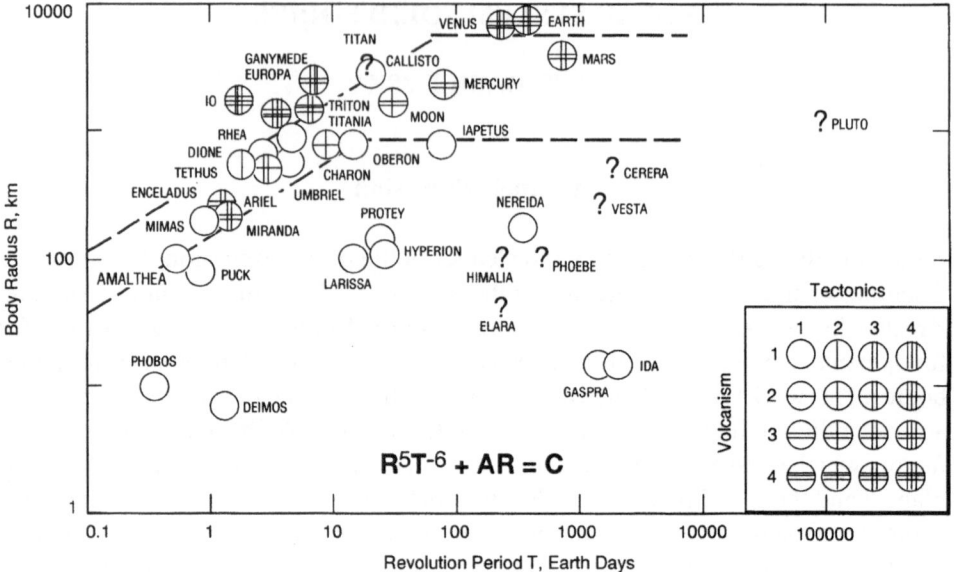

The field boundaries are specified by the equation AR+R5T-6=C, where A and C are constants chosen so that the boundaries best separate the objects with different degrees of endogenic activity.

This approach is the first approximation in the analysis; nevertheless, it provides a possibility to find and examine deviations from this correlation, and to make predictions of the endogenic activity of objects for which high-resolution images are not yet available. The deviations from the correlation are represented by four bodies: Callisto, Mimas, Umbriel, and Triton. 26 of the 30 bodies are consistent with this correlation. The reader is referred to the referenced paper for an explanation of the deviations.

More interesting now are predictions of late endogenic activity based on the body sizes and orbital periods. From their position on the diagram (shown by question marks), it is evident that even the largest asteroids (Ceres and Vesta) should not have late volcanic and tectonic resurfacing, and even less chance for other asteroids and small and distant satellites of the giant planets. Titan, the largest Saturnian satellite, has a good chance for significant late volcanism and tectonism. In the Pluto/Charon system, Charon has a better chance for late endogenic resurfacing than Pluto. The crucial issue of these predictions is the dynamic history of the bodies, which could be both more and less violent than their current orbital parameters would imply.

So, in keeping with the discussion guidelines, I should say that the above consideration gives high priority to missions that explore Titan, Pluto/Charon, and the largest asteroids.

References

1. Basilevsky, A.T., and Kreslavsky, M.A., "Volcanism and Tectonics on Planets and Satellites of the Solar System: Dependence on Size and Orbital Period." Astronomicheskii Vestnik, v.26, N2, March-April 1992, 66-76 (in Russian).

2. English translation of the paper in: Solar System Research, Consultant's Bureau, New York, September, 1992, 183-190.

GIANT PLANET FORMATION

A Comparative View of Gas-Accretion

G. WUCHTERL
Institut für Astronomie der Universität Wien
Türkenschanzstraße 17, A-1180 Wien, Austria
Internet: wuchterl@amok.ast.univie.ac.at

Abstract.
    The accumulation of giant planets involves processes typical for terrestrial planet formation as well as gasdynamic processes that were previously known only in stars. The condensible element cores of the gas-giants grow by solid body accretion while envelope formation is governed by 'stellar-like' equilibria and the dynamic departures thereof. Two hypotheses for forming Uranus/Neptune-type planets — at sufficiently large heliocentric distances while allowing accretion of massive gaseous envelopes, i.e. Jupiter-type planets at intermediate distances — have been worked out in detailed numerical calculations: (1) Hydrostatic gas-accretion models with time-dependent solid body accretion-rates show a slow-down of core-accretion at the appropriate masses of Uranus and Neptune. As a consequence, gas-accretion also stagnates and a window is opened for removing the solar nebula during a time of roughly constant envelope mass. (2) Gasdynamic calculations of envelope accretion for constant planetesimal accretion-rates show a dynamic transition to new envelope equilibria at the so called critical mass. For a wide range of solar nebula conditions the new envelopes have respective masses similar to those of Uranus and Neptune and are more tightly bound to the cores. The transitions occur under lower density conditions typical for the outer parts of the solar nebula, whereas for higher densities, i.e. closer to the Sun, gasdynamic envelope accretion sets in and is able to proceed to Jupiter-masses.

1. Introduction

The formation of giant planets will be discussed in the context of low-mass solar nebulae that are gravitationally stable (see Boss *et al.* 1989, for a

Earth, Moon, and Planets 67: 51–65, 1995.
© 1995 *Kluwer Academic Publishers.*

review of solar nebula models, Pollack 1985 for a discussion of the 'gas instability' hypothesis for giant planet formation in massive solar nebulae). Planetary growth in low mass solar nebulae first proceeds via solid body accretion (see Lissauer *et al.* 1995 for a review). When the growing planetary embryos are massive enough to gravitationally bind nebula gas, typically at $0.1M_\oplus$ ($M_\oplus$ is the Earth-mass) an envelope forms and gas accumulation starts. As the planetary embryos — cores of the protogiant planets — grow further, gas accretion becomes increasingly important and eventually has to exceed solid body accretion to form the massive gaseous envelopes of Jupiter and Saturn. This process of core-induced gas accretion will be considered in the following for non-rotating protoplanets in spherical symmetry.

For a discussion of slowly rotating protoplanets and their angular momentum evolutions in the quasi-spherical approximation, see Korycansky *et al.* (1991) for quasi-hydrostatic envelopes and Götz (1993) for hydrodynamics.

2. Equilibrium Envelopes around Accreting Cores

Early investigations of giant planet formation studied the equilibrium envelopes that are possible around growing condensible element cores embedded in the solar nebula (Perri and Cameron 1974, Mizuno, 1980, Stevenson 1982). The mechanical structure was obtained from hydrostatic equilibrium, the thermal structure from thermal equilibrium, i.e. energy losses of the envelope into the solar nebula are balanced by the energy input of incoming planetesimals, dissipating their kinetic energy upon impact. From a stellar structure point of view those envelopes are in 'complete' equilibrium, like, a main sequence star.

A key result was that beyond a certain 'critical' mass, no more envelopes in complete equilibrium could be found. This critical mass turned out to be independent of the density and temperature of the solar nebula as well as the distance from the Sun (Mizuno 1980). It depended, however on assumptions made for the dust opacity (Mizuno 1980), the mean molecular weight of the envelope and the core's planetesimal accretion rate (Stevenson 1982). For interstellar dust-opacities, and a constant core accretion rate of $10^{-6}M_\oplus/yr$ the critical core mass was found to be about $12M_\oplus$ in all cases, and therefore similar to the core masses inferred for the present giant planets. The critical envelope masses, however, fall short of Jupiter and Saturn, and are too large for Uranus and Neptune. Therefore the largest equilibrium envelopes did not correspond to the planetary envelope masses. A possible interpretation was, that Uranus and Neptune due to their longer core-accretion timescales did not reach their present, (and critical) core masses before the solar nebula was dispersed, whereas Jupiter and Saturn became critical and accreted to their present masses beyond the critical core mass.

Dynamical stability of the hydrostatic equilibrium was checked and found to hold at least up to the critical mass. This rules out the possibility of an envelope collapse initiated by small perturbations at the critical mass in the linear regime. Departures from thermal equilibrium were estimated to be small (Mizuno 1980).

The key reason for the non-dependence of protoplanetary structure on the solar nebula conditions is that their envelope structure rapidly approaches the radiative zero solution when integration proceeds from the outer boundary into the deeper atmospheric layers (Stevenson 1982). The conditions in the deep interior, that contains most of the envelope mass, are practically identical irrespective of atmospheric details. Stars lose this property when their atmospheres become convective (cf. e.g. Kippenhahn and Weigert 1990, p. 72ff). Motivated by a negative hydrodynamical result to form massive protogiant planets (Wuchterl 1991b), Wuchterl (1993) investigated the range of validity of the radiative zero solution for the planetary case and whether the transition from non-dependence to dependence on the outer boundary conditions might occur in the solar nebula. Envelopes with interstellar dust opacities and in complete equilibrium were investigated. The outer envelopes of protogiant planets became convective for nebula conditions with midplane densities only slightly higher than in the Kusaka et al. (1970) minimum reconstitutive mass solar nebula. Under these conditions protoplanetary structure and mass depend on the nebula conditions. Envelopes in complete equilibrium with masses up to $48M_\oplus$ were found. For a large range of envelope masses ($6 - 48M_\oplus$) the critical core mass changed only slightly. A family of protoplanets with a wide range of envelope masses at roughly constant core mass resulted from a set of different solar nebula densities. However, the largest envelope masses still fell short of Jupiter and Saturn.

The accuracy of masses obtained from these static solutions in complete equilibrium was checked by fully time-dependent radiation-hydrodynamical calculations. 'Dynamical' envelope masses at the static critical core masses were within 20% of the static values.

Either alternatives to the standard assumptions about dust-opacities and envelope compositions have to be invoked (Stevenson 1984, Lissauer et al. 1995) or there must be a departure from hydrostatic and/or thermal equilibrium during the early evolution of the giant planets to account for the sub-equilibrium envelope masses of Uranus and Neptune and the super-equilibrium masses of Jupiter and Saturn. Studies of departures from equilibrium are discussed in the remaining sections.

3. Quasi-Hydrostatic Protoplanets with Detailed Core-Accretion

A key question is the physical significance of the lack of solutions with equilibrium envelopes beyond the critical mass. This has to be investigated without using hydrostatic and/or thermal equilibrium.

The first truly time-dependent study of protoplanetary evolution was carried out by Bodenheimer and Pollack (1986). They still assumed the hydrostatic equilibrium to hold but allowed departures from thermal equilibrium by solving a time-dependent energy equation. This is analogous to investigating the quasi-hydrostatic pre-main sequence evolution of a young star. The non-existence of static protoplanetary envelopes beyond the critical mass might be caused by the energy input due to dissipation of planetesimal kinetic energy becoming insufficient to balance the losses into the ambient solar nebula. A protoplanetary envelope will then depart from thermal equilibrium and start to supply additional energy by quasi-hydrostatic contraction, i.e. gravitational compression of envelope gas. Bodenheimer and Pollack (1986) actually found the onset of contraction before the static critical core mass was reached. Envelope accretion is then controlled by the energy losses of the protoplanet into the solar nebula (cf. Bodenheimer and Pollack 1986). Accretion of nebula gas due to the envelope contraction exceeded the core accretion rate at about Mizuno's (1980) statical critical mass value for interstellar opacities and rapidly grew further. Gas accretion proceeded to Saturns mass, where it was shut off. The further evolution at constant mass was followed into the present, resulting in a mature Saturn-mass giant planet.

The rapid transition from low mass envelopes to efficient gas accretion left only a narrow time-window for removal of the solar nebula in order to stop envelope growth at masses appropriate for Uranus and Neptune.

Recently Pollack *et al.* (1995) refined the Bodenheimer and Pollack (1986) calculations. They calculated the core accretion rate in detail, using the 3-body cross sections for planetesimal collisions according to Greenzweig and Lissauer (1992). These rates are based on the concept of a single embryo accreting smaller 'background' planetesimals in a steady way. They cannot describe effects of radial migration in the planetesimal population and do not represent merging events (giant impacts). However, they are a major improvement to the constant values that were used in all previous studies of envelope accumulation.

Pollack *et al.* (1995) followed the evolution of core accretion together with quasi-hydrostatic evolution of the protoplanetary envelope. They found a stagnation of core-accretion after an early runaway growth. For an appropriate planetesimal surface density solid body accretion slows down for more then a million years at a large but subcritical core-mass. Gas accretion is very slow during this phase with the envelope being close to thermal equilibrium. The core and envelope mass increase slowly until the rising total mass extends the feeding zone of the protoplanet and core-accretion accelerates again. Envelope contraction becomes faster and the evolution can proceed to Saturn masses.

The essential difference to the earlier calculations with constant core accretion-rate is that the protoplanet spends substantial time in the stagnation phase with core and envelope masses comparable to Uranus' and Neptune's. This opens a wider window for the dispersal of the solar nebula at the appropriate masses. Jupiter and Saturn proceed beyond the stagnation phase. They enter the rapid gas accretion phase when the nebula is still present because the core accretion rates are higher close to the Sun.

In this picture the critical mass may be viewed as the hallmark of a breakdown of thermal equilibrium and the onset of quasi-hydrostatic contraction. Jupiter and Saturn become supercritical, start rapid but quasi-hydrostatic contraction and enter into a phase of efficient gas-accretion. Uranus and Neptune stay subcritical during the lifetime of the solar nebula and their envelope masses are 'frozen' to non-equilibrium values by the nebula-removal during a stagnation period of core accretion.

Since quasi-hydrostatic calculations apriori assume hydrostatic equilibrium it is interesting to investigate the stability of the envelopes obtained under this assumption, aposteriori. Bodenheimer and Pollack (1986) and Wuchterl (1991a) estimated the linear dynamical stability for adiabatic perturbations by using the envelope mass average of the first adiabatic exponent $\bar{\Gamma}_1$ as a rough stability indicator. Wuchterl 1991a, in a simplified analysis, discussed a stabilizing influence of the core decreasing the critical $\bar{\Gamma}_1$ below 4/3. Recently Tajima (1994), (see also Tajima and Nakagawa, this volume) calculated the quasi-hydrostatic evolution of a protoplanet from a subcritical total mass of $17 M_{\oplus}$ to $140 M_{\oplus}$. He carried out a detailed linear, adiabatic analysis of dynamical stability along with the evolutionary calculation. The hydrostatic equilibrium turned out to be stable throughout the calculation.

4. Hydrodynamics at the Static Critical Mass

Another possibility for investigation of the evolution of protogiant planets beyond the critical mass, i.e. beyond the largest known envelopes in hydrostatic and thermal equilibrium, is a fully time-dependent radiation-hydrodynamical approach (Wuchterl 1991a). The protoplanetary structure at the critical mass is then used as an initial condition. The further evolution is calculated without assuming hydrostatic or thermal equilibrium. A time dependent energy equation — as in quasi-hydrostatic studies — and

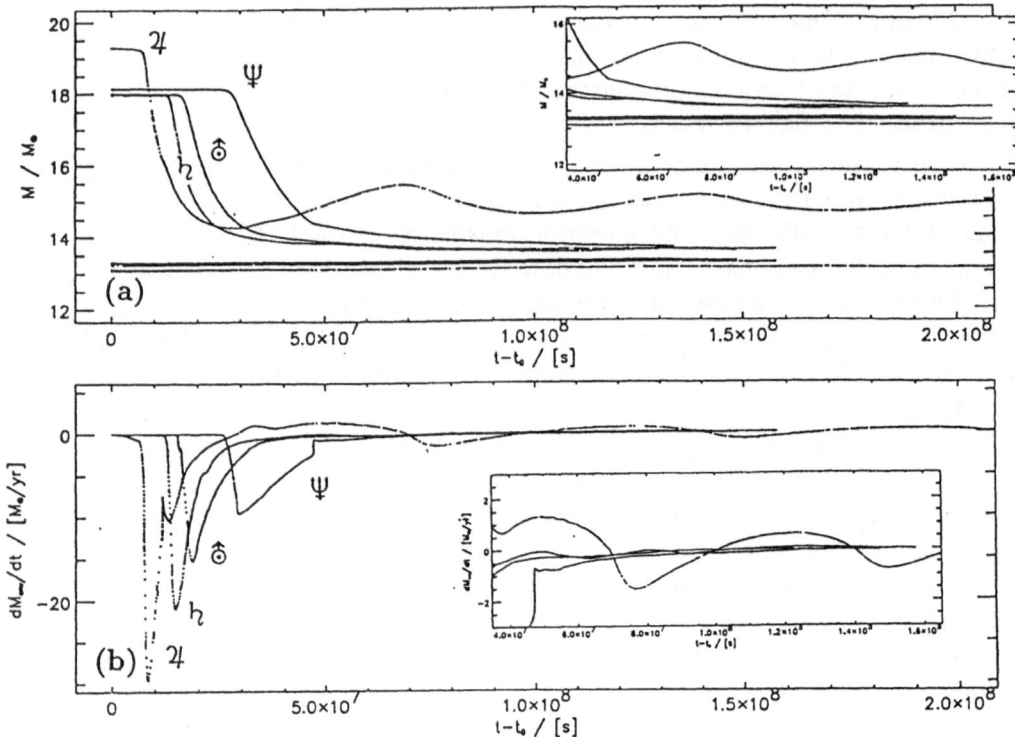

Figure 1. Envelope ejection at the critical mass. The evolution of protoplanets is calculated hydrodynamically with static critical mass models as initial conditions. Total mass and core mass, in Earth masses (top) and mass-flux in Earth masses per year (bottom) are shown as function of time in seconds. The planetary symbols denote solutions obtained for the locations of Jupiter, Saturn, Uranus and Neptune in the Kusaka *et al.* (1970) solar nebula. A large fraction of the envelope mass is lost and the mass flux approaches zero as a new 'post critical' equilibrium is approached in all cases. The core masses appear at constant values at the bottom of the top-figure.

the equation of motion for the envelope gas are solved. Solutions of this problem for a protoplanet at Jupiter's distance from the Sun were obtained by Wuchterl (1991b). Results for the Jupiter, Saturn, Uranus and Neptune positions in the Kusaka *et al.* (1970) solar nebula, for 'interstellar' dust opacities, and a constant core accretion-rate of $10^{-6} M_{\oplus}/\mathrm{yr}$ are shown in Fig. 1. Loss of a substantial fraction of the envelope mass on a dynamical timescale occurred in all cases. The mass-loss is driven by non-linear hydrodynamical waves. After the envelope mass is reduced, excitation of these waves fades and a new quasi-equilibrium state is approached. Core- and envelope masses of this new quasi-equilibrium are in all cases similar to Uranus' and Neptune's. A post-critical protoplanet has formed, with a critical core mass but a lower envelope mass.

After examination of the calculated flows and a local, linear, non-adiabatic stability analysis (Wuchterl 1990), Wuchterl (1991b) concluded that the excitation mechanism of the nonlinear waves is probably an opacity-mechanism, the so called κ-mechanism, that drives the stellar pulsations. The mass-loss at the critical mass, may thus be considered as analogous to pulsation-driven stellar winds. The κ-mechanism may well operate in dynamically stable systems (like stars) and is a kind of 'overstability', where the amplitudes of small, stable, adiabatic oscillations grow by a non-adiabatic effect. In case of the κ-mechanism amplitudes grow due to injection of energy by enhanced absorption of radiation at maximum compression and enhanced emission at maximum expansion.

After the dynamical stability estimates by Wuchterl (1991a), Bodenheimer and Pollack (1986) and the detailed linear, adiabatic analysis by Tajima (1994), a linear, non-adiabatic stability analysis could further clarify the linear dynamics of the envelope in the vicinity of the critical mass.

Long term integrations of the non-linear radiation-hydrodynamical problem show that post-critical protoplanets have a lower specific total energy than the critical initial state. The Uranus-Neptune-mass equilibria are therefore more tightly bound than a critical mass protoplanet. This indicates that a transition between the two states is energetically possible and the lower envelope-mass is energetically favored. The argument can be made without following the connecting dynamical evolution by comparing the initial and final equilibria alone.

5. A Quasi-Hydrostatic Comparison

Why do the quasi-hydrostatic investigations find accumulation to large envelope-masses (Bodenheimer and Pollack 1986, Pollack et al. 1995, Tajima 1994) while hydrodynamical studies (Wuchterl 1991b) show mass-loss at the critical mass?

A possible explanation would be, that the dynamical waves causing the mass-loss in hydrodynamical calculations are suppressed by assuming hydrostatic equilibrium in the quasi-hydrostatic calculations, much like a quasi-hydrostatic stellar-evolution calculation cannot show the dynamical stellar pulsations found, e.g. in the Cepheid instability strip. Reducing the equation of motion to hydrostatic equilibrium by dropping the 'inertial terms' should then produce agreement. Such a comparison was carried out by Bodenheimer, with the quasi-hydrostatic, 'Lagrangian' method used by Bodenheimer and Pollack (1986) and Pollack et al. (1995), and Wuchterl, using the hydrodynamical, 'self-adaptive grid' method described in Wuchterl (1991a). After eliminating differences in the model assump-

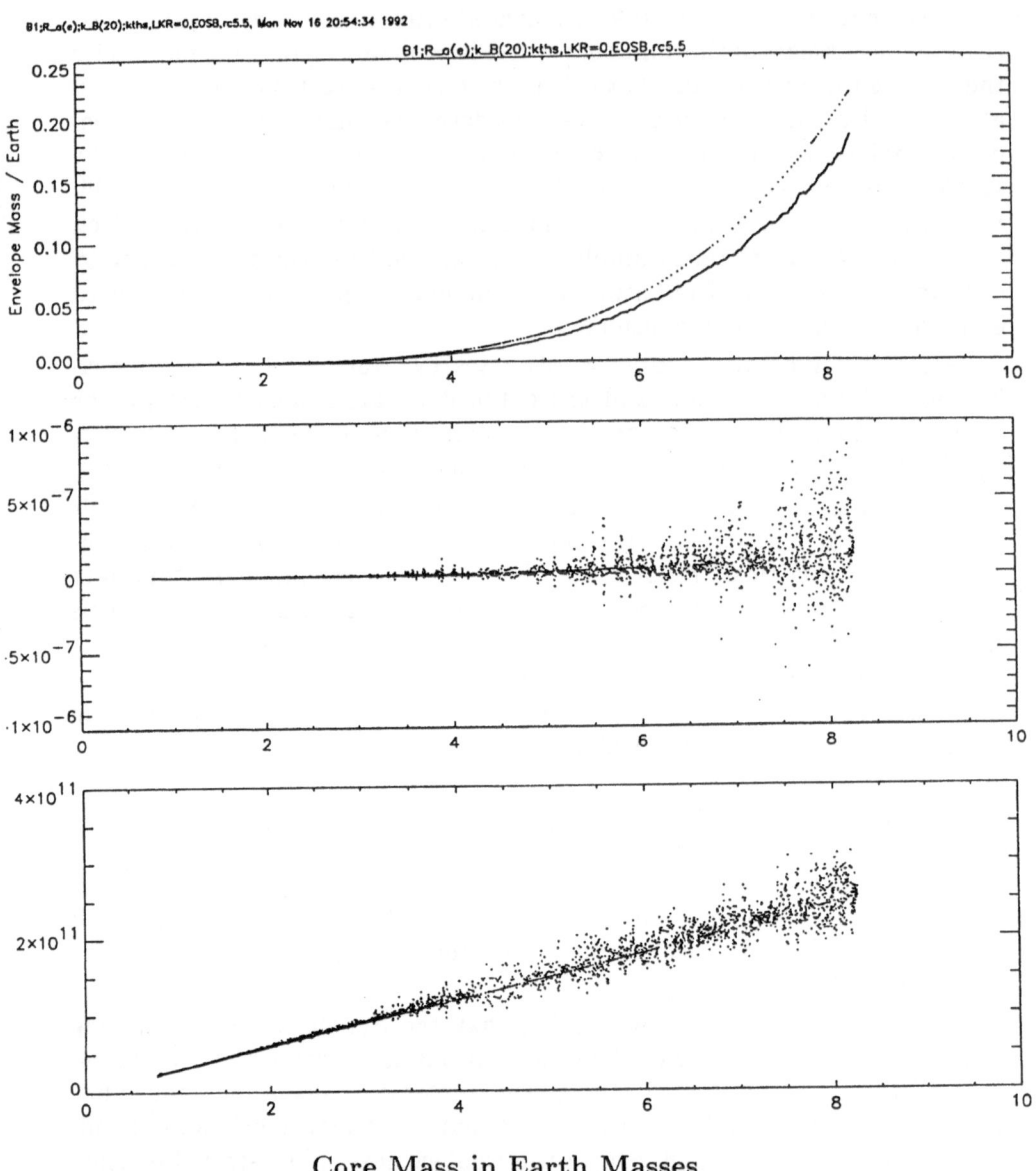

Core Mass in Earth Masses

Figure 2. Comparison of 'Lagrangian' and 'self-adaptive grid' calculations in the quasi-hydrostatic approximation. From top to bottom: envelope mass in Earth-masses, envelope mass-flux in Earth masses per year, the envelope radius in cm, are shown as a function of the core mass in Earth masses.

tions (equation of state, opacities and location of the outer boundary) calculations were carried out for the Bodenheimer and Pollack (1986), Case 1 parameters. The protoplanetary evolutions, when both calculations assume hydrostatic equilibrium, are shown in Fig 2. Overall there is good agreement. 'Lagrangian' envelope masses tend to be lower than 'adaptive-grid' masses. 'Lagrangian' envelope accretion-rates and radii show some scatter of probably numerical origin but agree on the average with the 'adaptive-grid' values that are visible as smooth curves within the scatter clouds. The envelope mass-averages of the first adiabatic exponent, a rough indicator for dynamical stability and the overall thermodynamical properties of the envelope-gas are also in detailed agreement.

In the core mass-range shown, both calculations also agree with a hydrodynamical calculation that would fall exactly on top of the quasi- hydrostatic 'adaptive grid' cures. Beyond $8M_{\oplus}$ the flow becomes dynamic in the hydrodynamical calculation, and the quasi-hydrostatic adaptive grid-calculation shows Mach-numbers approaching unity. The comparison at larger masses requires further discussion and is postponed to a future article.

The conclusion is, that in the quasi-hydrostatic limit the 'adaptive grid' and 'Lagrangian' method yield satisfactory agreement, when the hydrostatic equilibrium holds.

6. Hydrodynamics of 'No-Dust' Protoplanets

A remaining question is still how Jupiter, and massive envelopes in general can form when hydrodynamical effects are included. A key uncertainty in the formation calculations is the dust opacity. It has been assumed to be 'interstellar' in the preceeding discussion. Mizuno (1980) noticed a significant reduction of the critical core mass to about $1M_{\oplus}$, when no dust opacity is present. Stevenson (1982) found the critical mass to scale roughly as $\kappa^{3/7}$, where κ is the constant value of opacity in his analytical model. Bodenheimer and Pollack (1986) reduced the dust-opacity in their quasi-hydrostatical calculations by a factor of 50 and found a moderate increase in gas-accretion efficiency. To explore protoplanetary envelope-accretion for an extreme assumption concerning the dust opacity, the results of a hydrodynamical calculation with an opacity law of the form $\kappa = 10^{-8}P$, where P is the gas pressure and the constant is for cgs-units, are presented here. This opacity law has been proposed by Stevenson (cf. Lissauer et al. 1995) to model the opacity of pressure-induced absorption by molecular hydrogen. A comparison calculation with detailed zero-metallicity opacities shows no qualitative differences. Total (core plus envelope) mass accretion-rate is shown as function of the total mass for core accretion-rates of 10^{-8} and

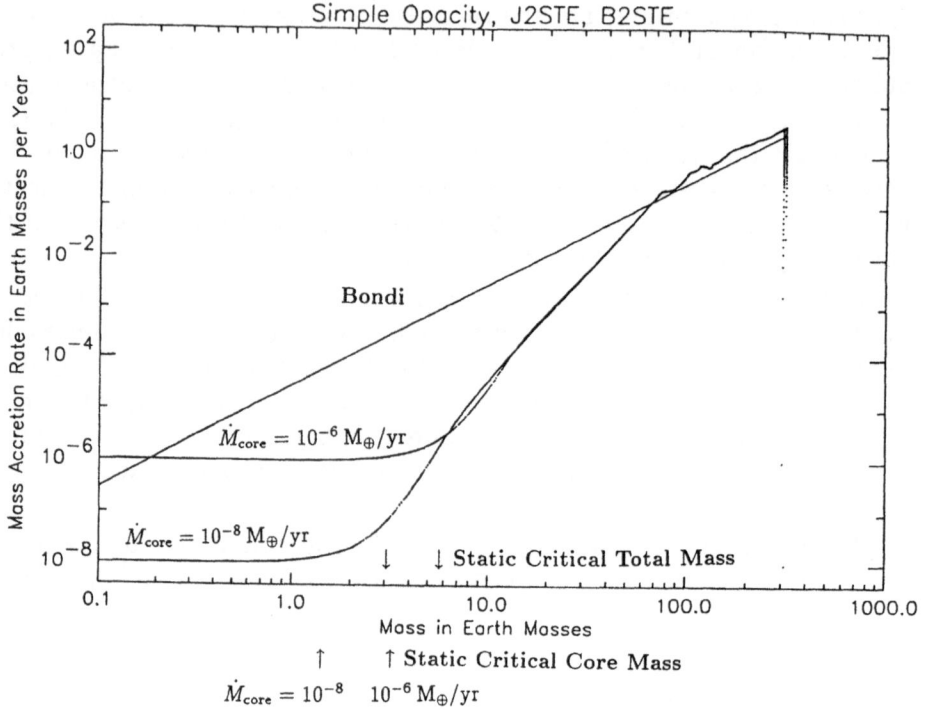

Figure 3. Hydrodynamics of 'No Dust'-protoplanets. Dynamical evolution calculations starting at small core masses are shown for a simple opacity model describing pressure induced absorption of molecular hydrogen. Total mass accretion rate (core plus envelope) in Earth masses per year is plotted as a function of total mass in Earth masses for core accretion rates of 10^{-6} (initially upper curve) and $10^{-8} M_\oplus/\text{yr}$. An estimate for the Bondi-accretion rate is plotted as a reference.

$10^{-6} M_\oplus/\text{yr}$. In both cases accretion proceeds to more than 300 $M_\oplus$ and a massive protoplanetary envelope is formed. The envelope contraction starts quasi-hydrostatically, gradually accelerates into the dynamic regime and passes Mach 0.01 at about 50 $M_\oplus$. At about 300 $M_\oplus$ the envelope collapses overall and the flow is partially supersonic. Note that it has been assumed that the solar nebula is an unlimited reservoir of gas and effects of tidal truncation have been neglected.

The mass ejection found for 'interstellar' dust opacities does not occur. Accretion proceeds to Jupiter's mass where the envelope collapses.

7. Hydrodynamics of Largely Convective Envelopes

Is it possible to form massive 'dusty' envelopes without assuming hydro-static equilibrium to avoid the ejection at the critical mass? Standard pro-toplanets with radiative outer envelopes have similar structure (Mizuno 1980, Stevenson 1982) and similar hydrodynamical behaviour (cf. Fig. 1) at the critical mass, for a wide range of solar nebula conditions. Candidates for a different dynamical behaviour are the largely convective protoplanets (Wuchterl 1993 and Sect. 2). Their outermost envelopes are convective and as a consequence show different static structure for different nebula con-ditions. Their static envelopes are substantially larger than the standard 'radiative-zero'-envelopes.

Hydrodynamical evolutions for a sequence of nebula-conditions yield-ing protoplanets ranging from radiative to largely convective are shown in Fig. 4. The radiation hydrodynamical equations described by Wuchterl (1991a) were solved, starting at very small core mass (core size about 10 km). Interstellar dust opacities and a constant core accretion-rate of $10^{-6} M_\oplus/\text{yr}$ are assumed. The five evolutionary curves in Fig. 4 are ob-tained for a set of solar nebula conditions with nebula density increasing from the left curve to the right. The leftmost evolutionary track is ob-tained for the standard Kusaka et $al.$, 'Jupiter' conditions. It is the same that Mizuno (1980) used in his static studies and is close to Bodenheimer and Pollacks (1986), quasi-hydrostatic Case 2. These nebula conditions are also identical to those used for the hydrodynamics starting at the static critical mass, cf. Fig. 1 and allow direct comparison of hydrodynamics with critical mass and 'low mass' initial conditions.

For the standard Jupiter-case with 'low-mass' initial conditions the typ-ical accretion behaviour is seen in Fig. 4 up to the static critical core mass (about $13 M_\oplus$). Static and quasi-hydrostatic studies are confirmed by hy-drodynamics. At a core mass close to $15 M_\oplus$ the evolutionary track turns sharply and proceeds horizontally to the left in the total-mass core-mass diagram. This is due to envelope ejection reducing the total mass on a dy-namical timescale, and therefore at constant core-mass. The evolution is followed until the post-critical quasi-equilibrium is reached (left end of the track). The four tracks to the right are obtained for the same assumptions but the solar nebula density is enhanced from $1.5\,10^{-10}\text{g/cm}^3$, ('Jupiter') to $3, 6, 10$ and $13\,10^{-10}\text{g/cm}^3$, i.e. cases JR3, JR6, JR10, JR13, respectively. Note that neighbouring tracks differ by only about a factor two in nebula density. At subcritical core-masses, the effect known from the static study is seen. Convective outer envelopes have larger masses at a given core mass (Wuchterl 1993). But envelope ejection and transition to post-critical equi-libria is still found for cases JR3 and JR6. At $10\,10^{-10}\text{g/cm}^3$, JR10, roughly

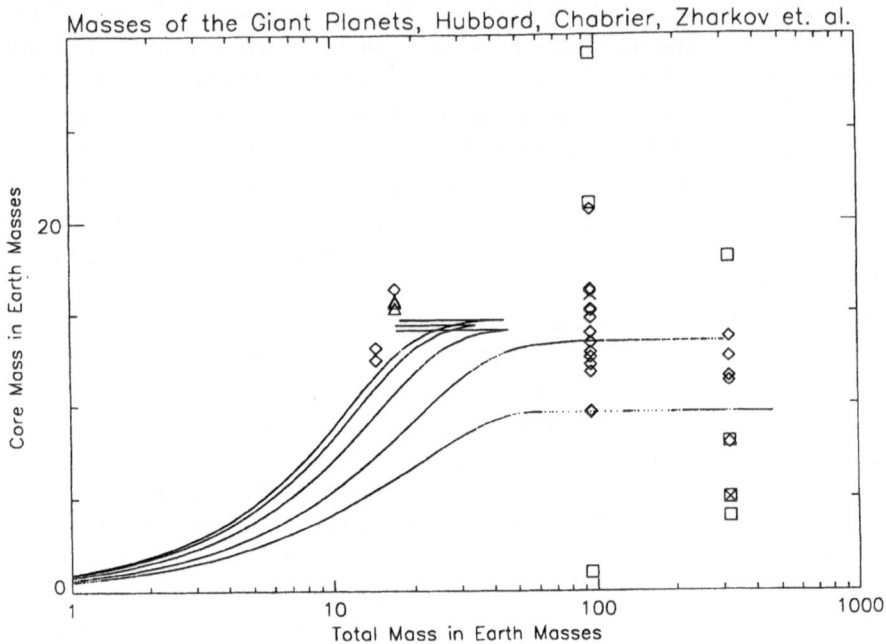

Figure 4. Hydrodynamics starting at low core mass for a set of nebula conditions with increasing solar nebula density. Core mass vs. total mass, both in earth masses. Superimposed are the core masses inferred for the present giant planets by optimized interior models according to Hubbard and Marley (1989), Hubbard *et al.* (1991), Chabrier *et al.* (1992) and Zharkov and Gudkova (1991, 1992).

a factor 8 above the standard density, a qualitatively different behaviour occurs. Hydrodynamical waves are excited but now cannot drive mass-loss — no envelope ejection occurs. The protoplanet enters a phase of increasingly rapid gas accretion. The evolutionary track continues to large envelope and total masses. The late phases of accretion may be modified by the effects of limited gas supply from the nebula and/or tidal truncation. Both have been neglected here. Case JR13, the rightmost track shows onset of gas accretion at even lower core mass. The evolutions of JR10 and JR13 are followed until their envelopes enter supersonic collapse. Note that the core mass at the onset of rapid gas-accretion is only slightly smaller than 'radiative-zero' critical core masses.

Superimposed on the dynamical evolutionary tracks are condensible element mass values ('core'-masses) inferred for the present giant planets (see Hubbard, this volume, for a discussion). Protoplanets for the inferred range of core- and total masses can be obtained from the nebula density-sequence discussed above. Below a certain density *all* protoplanets (for interstellar dust-opacity and a given core accretion-rate) eject their envelopes. This is a consequence of the radiative structure of their outer envelopes leading to the same *dynamical* behaviour, independent of the nebula conditions (cf. Fig. 1). Only when the envelopes become sufficiently convective at higher nebula densities the ejection mechanism is damped, and accretion can proceed to large envelope masses. This may be viewed as analogous to the 'red edge' of the Cepheid instability strip in the Hertzsprung-Russel-diagram, where convection plays a key role in damping stellar pulsations.

Formation of massive 'dusty' protoplanetary envelopes occurs at higher nebula densities and is therefore favored closer to the Sun.

8. Conclusion

Formation of protogiant planets with masses corresponding to the present giant planets can be understood on the basis of core-accretion with hydrostatic envelopes plus removal of the solar nebula and/or as a consequence of envelope hydrodynamics. A further comparative approach and detailed examination of the properties of the present giant planets will be needed to 'break the weak parts' of formation theories. An interesting step toward a prediction of extrasolar planetary masses is a consequence of the hydrodynamical studies: Uranus-Neptune-type giant planets are a standard result for a very wide range of solar nebula conditions. Formation of massive Jupiter-type planets requires special, i.e. dense nebula conditions to make the envelopes sufficiently convective.

9. Acknowledgements

I thank Peter Bodenheimer for the intense collaboration during the ITP-program on planet formation at UCSB, where the quasi-hydrostatic comparison was calculated. This work was supported in part by the National Science Foundation under Grant PHY89-04035 and the German Deutsche Forschungsgemeinschaft, DFG, Project No. Ts 17/2-2,3. Work for this article was supported in part by the Austrian Fonds zur Förderung Wissenschaftlicher Forschung, FWF under project No. P8806-PHY. I thank David Stevenson for his review-comments.

10. References

Bodenheimer, P. and Pollack, J.B.(1986) Calculations of the accretion and evolution of the giant planets: The effect of solid cores, *Icarus* **67**, 391–408.

Boss, A.P., Morfill, G.E., and Tscharnuter, W.M. (1989) Models of the Formation and Evolution of the Solar Nebula, in S.K. Atreya, J.B. Pollack and M.S. Matthews, (eds.), *Origin and Evolution of Planetary and Satellite Atmospheres*, Univ. of Arizona Press, Tucson, pp. 35–77.

Chabrier, G., Saumon, D., Hubbard, W.D., and Lunine, J.I. (1992) The Molecular-Metallic Transition of Hydrogen and the Structure of Jupiter and Saturn, *Astrophys. Journ.* **319**, 817–826.

Greenzweig, Y. and Lissauer, J.J.(1990) Accretion Rates of Protoplanets II. Gaussian Distribution of Planetesimal Velocities, *Icarus* **100**, 440–463.

Götz, M. (1993) *Die Entwicklung von Proto-Gasplaneten mit Drehimpuls — Strahlungshydrodynamische Rechnungen*, Dissertation, Univ. Heidelberg.

Hubbard, W.B. and Marley, M.S. (1989) Optimized Jupiter, Saturn and Uranus Interior Models, *Icarus* **78**, 102–118.

Hubbard, W.B., Nellis, W.J., Mitchell, A.C., Holmes, N.C., Limaye, S.S., and McCandless, P.C.(1991) Interior Structure of Neptune: Comparison with Uranus, *Science* **253**, 648–651.

Kippenhahn, R. and Weigert, A. (1990) *Stellar Structure and Evolution*, Springer-Verlag, Berlin.

Korycansky, D.G., Bodenheimer, P., and Pollack, J.B.(1991) Numerical models of giant planet formation with rotation, *Icarus* **92**, 234–251.

Kusaka, T., Nakano, T. and Hayashi, C. (1970) Growth of Solid Particles in the Primordial Solar Nebula, *Prog. Theor. Phys.* **44**, 1580–1595.

Lissauer, J.J., Pollack, J.B., Wetherill, G.W. and Stevenson, D.J., (1995) Formation of the Neptune System, in *Neptune* Univ. Arizona Press, in prep.

Mizuno, H.(1980) Formation of the Giant Planets, *Prog. Theor. Phys.* **64**, 544–557.

Perri, F. and Cameron, A.G.W. (1974) Hydrodynamic instability of the solar nebula in the presence of a planetary core, *Icarus* **22**, 416–425.

Pollack, J.B.(1985) Formation of the giant planets and their satellite-ring systems: An overview, in D.C. Black and M.S. Matthews (eds.) *Protostars and Planets II*, Univ. Arizona Press, Tucson.

Stevenson, D.J.(1982) Formation of the giant planets, *Planet. Space Sci.* **30**, 755–764.

Stevenson, D.J.(1984) On forming giant planets quickly (superganymedean puffballs!), *Lunar Planet. Sci.* XV, 821–822 (abstract).

Tajima, N.(1994) *Giant Planet Formation: Dynamical Stability of the Envelope*, Master Thesis, Univ. Tokyo.

Wuchterl, G.(1990) Hydrodynamics of Giant Planet Formation I: Overviewing the κ-Mechanism *Astron. Astrophys.*, **238**:83-94.

Wuchterl, G.(1991a) Hydrodynamics of Giant Planet Formation II: Model Equations and Critical Mass *Icarus* **91**, 39–52.

Wuchterl, G.(1991b) Hydrodynamics of Giant Planet Formation III: Jupiter's Nucleated Instability"'.
Icarus **91**, 53–64.

Wuchterl, G., 1993
The Critical Mass for Protoplanets Revisited: Massive Envelopes Through Convection"'.
Icarus **106**, 323–334.

Zharkov, V.N. and Gudkova, T.V. (1991) *Ann. Geophys.* **9**, 357.

Zharkov, V.N. and Gudkova, T.V. (1992) Modern Models of Giant Planets, in Y. Soyono and M.H. Manghnani (eds.), *High Pressure Research: Application to Earth and Planetary Sciences*, Terra Sci. Publ. Co. (TERRAPUB), Tokyo / American Geophysical Union, Washington, D.C..

GIANT PLANET FORMATION
Dynamical Stability of a Massive Envelope

NOBUYA TAJIMA
Department of Earth and Planetary Physics, University of Tokyo
University of Tokyo, Bunkyo-ku, Tokyo 113, Japan
E-mail : tajima@geoph.s.u-tokyo.ac.jp

YOSHITSUGU NAKAGAWA
Department of Earth and Planetary Sciences, Kobe University
Kobe University, Nada-ku, Hyogo 657, Japan
E-mail : yoshi@saturn.phys.kobe-u.ac.jp

There are two antithetic works on the formation stage of giant planets. Bodenheimer and Pollack (1986) showed with numerical calculation that giant planets are formed through quasi-static contraction of massive envelopes. On the other hand, Wuchterl (1991) obtained the results that giant planets should undergo a dynamical phase and cannot have massive envelopes like the present Jupiter and Saturn because dynamical oscillation would eject envelope mass.

In order to understand giant planet formation processes, we studied the structure and evolution of the accumulating envelope, and checked the dynamical stability of the obtained structure against a linear adiabatic perturbation. We calculated the structure and evolution of that by means of the Entropy Method (Tajima, 1994). The concept of the Entropy Method is to calculate dynamical structure and energy transfer alternatively. Roughly speaking, from the entropy distribution at each time step, we numerically obtain a hydrostatic structure and check the stability of the envelope. Then, if the obtained envelope is dynamically stable, the evolved entropy distribution is calculated by solving the equation of energy transfer. The numerical procedure of checking dynamical stability is similar to Mizuno *et al.* (1978).

In our calculation, we assume that the envelope evolves quasi-statically. The equation of state include effects of hydrogen molecular dissociation and hydrogen and helium first ionization. But it does not include non-ideal effects at high pressure. We assume that the whole envelope is always an ideal gas. Then the comparison is not strictly valid. However, during our calculation, such a high pressure region have quite a little mass and contribution of non-ideal effects to global structure of envelope is considered to be small. In order to calculate structure and evolution

Earth, Moon, and Planets 67: 67–69, 1995.
© 1995 *Kluwer Academic Publishers.*

of much more massive envelope, it is true that we must take account of non-ideal effects. We use the same opacity data as that of Bodenheimer and Pollack (1986).

To get an initial state, we calculated steady state structure in hydrostatic equilibrium and studied its dynamical stability. The relation between the total mass and core mass is shown in Figure 1. Open circles denote the dynamically stable states and solid circles denote the unstable states. As Mizuno *et al.* (1978) and Mizuno (1980) showed, we also found some structures, which is dynamically unstable for linear and adiabatic perturbation. For an initial state of our evolutionary calculation, we adopted the "critical state" that the core mass is $12.5M_E$ (where M_E is the mass of the Earth) and total mass is about $19M_E$. This initial state is similar to that of Wuchterl (1991).

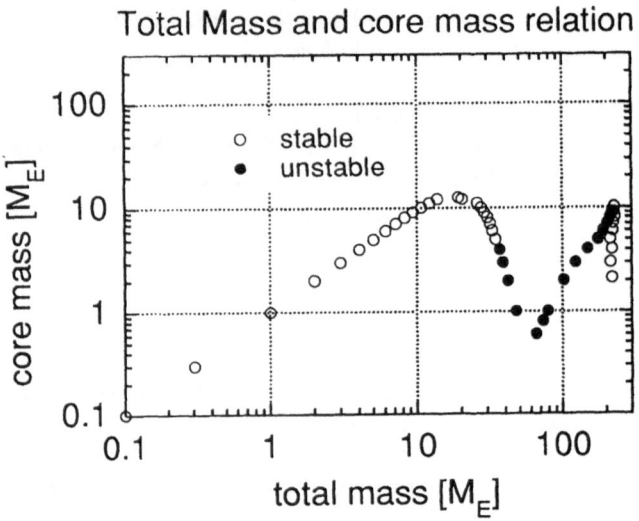

FIGURE 1. The total mass and core mass relation in hydrostatic and steady state. Open circles are dynamically stable and solid circles are unstable for linear and adiabatic perturbation.

The outer boundary conditions, in our evolutionary calculation, are imposed at the point r_{out} which is far from the Hill radius. At the point r_{out}, we put the temperature $T = 123$ K and the density $\rho = 1.5 \times 10^{-11}$ g cm^{-3} for Jovian region after Hayashi (1981). We set the region between Hill radius and r_{out} non-self-gravity. The thermodynamical quantities at the Hill radius are almost the same as those at the point r_{out}. The reasons why we put such a region are to get rid of effects of the outer boundary conditions from the dynamical stability of the envelope and also from gas accretion process.

In Table 1, we summerize the points of difference among our work, Bodenheimer and Pollack (1986), and Wuchterl (1991). The gas mass means the mass inside Hill sphere excluding core mass. Our calculation code is completely independent of that of Bodenheimer and Pollack (1986). Furthermore, insure the validity of the assumption of quasi-static evolution, we analyzed the dynamical stability. As a result, we did not find any instability during the accumulation until mass of the protoplanet exceeds $79M_E$. Our results confirm the work of Bodenheimer and Pollack (1986). Our dynamical stability check is valid for linear and adiabatic perturbation, and there will be a possibility that the hydrodynamical wave is excited by non-linear and non-adiabatic modes. (Wuchterl, personal communication.) Another possibility is that Wuchterl's calculation became dynamically unstable due to his outer boundary condition. Furthermore, the difference of adopted opacity data might affect the dynamical stability.

TABLE 1. The points of difference among our work,
Bodenheimer and Pollack (1986), and Wuchterl (1991a,b)

	Bodenheimer and Pollack (1986)	Wuchterl (1991a,b)	Our Work
evolutionary code	Quasi-static	Dynamical	Quasi-static
stability analysis	No		Yes
core accretion rate	10^{-5}	10^{-6}	10^{-6}
(M_E/year)	constant	constant	constant
final core mass (M_E)	29	13.1	12.8
final gas mass (M_E)	66	1.4	66.2
evolution	Quasi-static	Quasi-static	Quasi-static
(contraction)		$\Rightarrow$ Dynamical Oscillation	Any time stable

Acknowledgments

We wish to thank Prof. P. Bodenheimer very much for giving us his opacity data. We also thank G. Wuchterl for fruitful discussion. We also acknowledge S. Sasaki for useful advice and fruitful discussion. Numerical simulations were done by the HITACHI 3050RX at Department of Earth and Planetary Physics, the University of Tokyo.

References

Bodenheimer, P. and Pollack, J. B. (1986). Calculations of the Accretion and Evolution of Giant Planets: The effects of Solid Cores *ICARUS*. **67**, 391-408.

Hayashi, C. (1981). Structure of the Solar Nebula, Growth and Decay of Magnetic Fields and Effects of Magnetic and Turbulent Viscosities on the Nebula. *Prog. Theor. Phys.* **70**, 35-53.

Mizuno, H. (1980). Formation of the Giant Planet *Prog. Theor. Phys.* **64**, 544-557.

Mizuno, H., Nakazawa, K. and Hayashi, C. (1978). Instability of a Gaseous Envelope Surrounding a Planetary Core and Formation of Giant Planet. *Prog. Theor. Phys.* **60**, 699-710.

Tajima, N. (1994) *Giant Planet Formation: The Dynamical Stability of the Envelope.*, Master Thesis, The University of Tokyo.

Wuchterl, G. (1991a) Hydrodynamics of Giant Planet Formation II: Model Equations and Critical Mass *ICARUS*. **91**, 39-52.

Wuchterl, G. (1991b) Hydrodynamics of Giant Planet Formation III: Jupiter's Nucleated Instability *ICARUS*. **91**, 53-64.

ORIGIN OF THE MAJOR PLANET ATMOSPHERES: CLUES FROM TRACE SPECIES

S.K. ATREYA and S.G. EDGINGTON
Department of Atmospheric, Oceanic and Space Sciences
The University of Michigan, Ann Arbor, MI 48109-2143

D. GAUTIER
Observatoire de Paris-Meudon, France

T.C. OWEN
Institute for Astronomy, University of Hawaii
Honolulu, HI 96822

According to the generally accepted planetesimal accretion hypothesis, the present atmospheres on the major planets, Jupiter, Saturn, Uranus and Neptune, resulted from two sources: outgassing from planetary interiors during the accretionary heating phase - the planetesimals making up the core were composed of grains, ices and possibly clathrates - and direct capture of gases from the solar nebula. It is not apparent, however, whether the present atmospheres of these planets are representative of the composition of the dense interstellar cloud from whose collapse the solar nebula initially formed, or whether the interstellar material was further processed in the solar nebula before being accreted into the major planets. Although present observations cannot fully constrain the models of the origin of atmospheres, clues to the physical and chemical nature of the solar nebula and subsequent origin and evolution of the major planet atmospheres are beginning to emerge from studies of the elemental isotopic ratios (such as D/H) and the mole fractions of such trace species as He, CH_4, NH_3, CO and HCN. Since much has been written on D/H, He/H_2 and CH_4/H_2 [1], this paper focuses on what one can learn about the origin of the major planet atmospheres from the detected presence of CO and HCN in Neptune's atmosphere. On the basis of our calculations, we conclude that the interior of Neptune should contain CO and N_2, with the N_2 mole fraction close to the value for the cosmic C/N ratio. The presence of these interstellar molecules in the interior of Uranus is suggested also.

CO and HCN have been detected at ppm and ppb levels, respectively, in the stratosphere of Neptune [2,3] and their tentative implications for atmospheric chemistry have been discussed [2,4,5,6,7]. Extraplanetary sources of water vapor (e.g. from ablating meteoroids or comets) and nitrogen (e.g. from Triton) could, in principle, produce CO and HCN following their reactions with the photolysis products of CH_4; they are, however, inadequate for producing the observed mole fractions of CO and HCN [2,4,5,6,7]. The alternative is an intrinsic source in the form of CO and N_2. This is attractive from many different points of view: (a) it does not require a unique origin for Neptune's atmosphere relative to Uranus' - a planet that is similar to Neptune in "bulk" composition - since the lack of detection of CO and HCN on Uranus can be attributed to the lack or virtual lack of convection in Uranus' interior which would prevent detectable abundances of CO and N_2 from being transported to the stratosphere (presence of CH_4 there could create a potential problem, but it probably has another explanation, as discussed later); (b) it reduces the value of the helium mole fraction on Neptune to 15% [8], which is consistent with the protosolar and the Uranian values, provided that the N_2 mole fraction is 0.3% (N_2 of 0.3% is what one would expect for

Earth, Moon, and Planets 67: 71–75, 1995.
© 1995 *Kluwer Academic Publishers.*

solar C/N ratio with C/H = 30x solar); (c) it explains the observed depletion of NH_3 on Neptune (and Uranus, if $C/N \geq$ solar) since most of the nitrogen would be in the form of N_2. The proposed presence of CO and N_2 in the interiors of Uranus and Neptune implies that these species were incorporated directly from the solar nebula, and that the N_2 and CO of the interstellar cloud were not reduced to NH_3 and CH_4, respectively, in the less than 30 Myr lifetime of the solar nebula before its gas was blown away. Indeed, it seems that equilibrium may not have reached anywhere in the solar nebula despite the possibility of Fischer-Tropsch-type catalytic reactions, since the pressures and temperatures in the solar nebula were inadequate for such reactions. The fact that both Jupiter and Saturn have solar or super-solar NH_3 and CH_4 in their atmospheres appears to imply that once incorporated as N_2 and CO from the solar nebula these gases must have been reduced to NH_3 and CH_4, respectively, by Haber-type processes in the deep, dense, hot interiors of these planets. Such reduction requires presence of an iron catalyst which is expected to be present in their cores. Although the cores of Uranus and Neptune are also expected to have iron, the gases could be separated from the iron by layering in the interior. Since the CO to CH_4 catalysis proceeds somewhat differently than N_2 to NH_3, there is the possibility that some CO hydrogenation occurred in the solar nebula, but most of the CO was reduced to CH_4 in the interiors of Uranus and Neptune as well, leaving some CO behind. It is this CO that is detected in the atmosphere of Neptune, as well as on Jupiter and Saturn. The detection of N_2 and CO on Pluto and Triton lends support to this hypothesis, since the Haber-type reactions cannot occur on these tiny solid bodies, so that they are expected to retain to a large extent the original gases incorporated during their accretion.

Regarding the large CH_4 abundance on the major planets, there is yet another likelihood for the formation of methane on these planets. Since most of the carbon in the interstellar cloud is in the form of grains, perhaps carbon delivered as grains was responsible for the formation of methane.

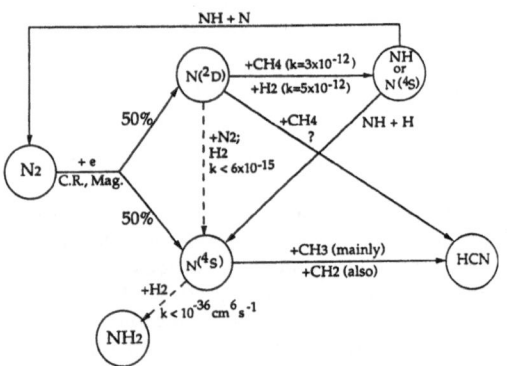

Figure 1. Nitrogen chemistry leading to the formation of HCN.

Dissociation of N_2 in the stratosphere, after this gas convects from Neptune's interior, is necessary for producing HCN as detailed in Figure 1 (intermediate product H_2CN is not shown). The dissociating/ionizing flux is most likely provided by the galactic cosmic rays (GCR). The production rate of ion-electron pairs created in gas j by primary GCR [9] can be approximated by (assuming that energy is deposited locally):

$$q_j(z) = \frac{1}{\Delta E_j} \int_\Omega \int_{E_0}^\infty \Phi(E, z, \Omega)\left[-\frac{dE}{dz}\right]_j dEd\Omega \tag{1}$$

where ΔE_j is the average energy needed to create an ion-electron pair in gas j and is typically 36 eV, $[-dE/dz]_j = n_j\sigma_j E$ is the stopping power, Ω is the solid angle, E is the energy of the cosmic ray, and $\Phi(E, z, \Omega)$ is the differential flux of GCR. The total absorption cross-section for the incident galactic cosmic rays in the atmosphere of Neptune is taken to be $8.5 \times 10^{-26} cm^2$, and it is based on the work of Schopper [10] and Belletini et al [11]. Secondary electrons, produced in the collision of galactic cosmic rays with the atmospheric constituents, dissociate and ionize N_2 and H_2 in the following manner:

$$
\begin{array}{llll}
e + N_2 \rightarrow N_2^+ + 2e & q = 0.53 & \sigma = 2.6 \times 10^{-16} cm^2 & (2) \\
\quad\quad \rightarrow N(^4S) + N^+ + 2e & = 0.12 & = 6.0 \times 10^{-17} cm^2 & (3) \\
\quad\quad \rightarrow N(^4S) + N(^2D) + e & = 0.35 & = 1.8 \times 10^{-16} cm^2 & (4)
\end{array}
$$

and

$$
\begin{array}{lll}
e + H_2 \rightarrow H_2^+ + 2e & \sigma = 1.0 \times 10^{-16} cm^2 & (5) \\
\quad\quad \rightarrow H^+ + H + e & = 6.0 \times 10^{-18} cm^2 & (6)
\end{array}
$$

where q is the quantum efficiency and σ is the cross section for a given path [12,13]. The previous values of q's for reactions (2), (3) and (4), 0.45, 0.10 and 0.45, respectively, do not alter the results significantly.

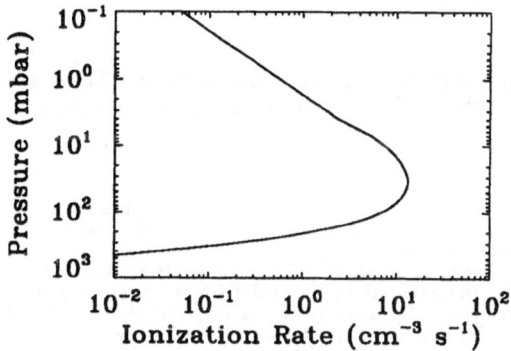

Figure 2: Ion-pair production rate due to penetration of primary galactic cosmic rays into Neptune's atmosphere at a magnetic latitude of 60°.

Using the differential flux of the GCR at Neptune [14], we obtain total ion-pair production rate shown in Figure 2 and a column production rate of ion pairs, $q_T = 6.7 \times 10^7 cm^{-2}s^{-1}$. The $N(^4S)$ production rate is 73% of the total nitrogen ion production rate, $q_{N_2^+}$; whereas $q_{N_2^+} = 3.3 q_{H_2^+} f_{N_2}$, where $q_{H_2^+}$ is the total hydrogen ion (H_2^+ and H^+) production rate and f_{N_2} is the N_2 mole fraction. Since $q_{H_2^+} \approx q_T$, the $N(^4S)$ production is $1.6 \times 10^8 f_{N_2}$. The $N(^4S)$ required to produce the observed ppb level of HCN is $1.7 \times 10^6 cm^{-2}s^{-1}$ for $K = 10^5 cm^2 s^{-1}$ [2] in the production region of HCN (1-2mb) and $1.7 \times 10^5 cm^{-2}s^{-1}$ if $K = 10^4 cm^2 s^{-1}$, where K is the eddy diffusion coefficient. Therefore 0.1%-1% mole fraction of N_2 will be needed. The actual required f_{N_2} is probably closer to 0.1% than 1% since $K < 10^5 cm^2 s^{-1}$, $N(^2D)$ could quench and produce 100% more $N(^4S)$, and the actual cascading process of the charged particles through the atmosphere than the simplified scenario of local deposition of energy considered here could produce 50-100% more $N(^4S)$. Proper treatment of the cascading of charged particles through the atmosphere requires a Monte Carlo-type approach where the history of cosmic ray particles and secondary electrons through the atmosphere is followed.

In summary, we propose that the major planets acquired at least a part of their atmospheres from vaporization of ices embedded in the nebula. Whatever the form of C and N compounds in these ices, the material was reprocessed in the planetary interiors. A possible exception to this scenario is provided by laboratory studies of the trapping of gases by ices at low temperature [15]. This work indicates that the relative fractions, as well as abundances of CO and N_2 (as well as other volatiles) trapped in ices, has a strong dependence on temperature. If we believe that these volatiles were delivered to Uranus and to Neptune (and other major planets) by the icy planetesimals, then there is the possibility that far less CO and N_2 were delivered to Uranus than Neptune, which formed in the colder part of the solar nebula [16].

Observational tests as well as improved theoretical models will be needed to confirm and improve upon the hypothesis proposed here. Direct spectroscopic detection of N_2 on the major planets is not possible at this time. However, detection and quantification of nitrogen bearing molecules, such as nitriles, as well as C/H (as in CH_4), P/H (as in PH_3) and the isotopic ratios in the noble gases are relevant and crucial. Thermochemical models for the interior and models of the chemistry initiated by the solar ultraviolet photons and charged particles, including Monte Carlo calculations for determining the N-atom yield are essential. Comparative planetology in this instance has the potential of addressing some of the most fundamental questions of the formation of the planets and their atmospheres.

Acknowledgment
S.K. Atreya acknowledges support received from the NASA Solar System Exploration Division, Planetary Atmospheres Program for research reported here.

REFERENCES

1. Gautier, D. and Owen, T.C. (1989) The composition of the outer planet atmospheres, in S.K. Atreya, J.B. Pollack and M.S. Matthews (eds.),*Origin and Evolution of Planetary and Satellite Atmospheres* , University of Arizona Press, Tucson, pp. 487-512.
2. Marten, A., Gautier, D., Owen, T., Sanders, D.B., Matthews, H.E., Atreya, S.K., Tilanus, R.P.J. and Deane, J.R. (1993) First observations of CO and HCN on Neptune and Uranus at millimeter wavelengths and their implications for atmospheric chemistry, *Astrophys. J.* **406**, 285-297.
3. Rosenqvist, J., Lellouch, E., Romani, P.N., Paubert, G. and Encrenaz, T. (1992) Millimeter-wave observations of Saturn, Uranus, and Neptune: CO and HCN on Neptune, *Astrophys. J.* **392**, L99-L102.
4. Atreya, S.K., Owen, T.C., Gautier, D. and Marten, A. (1992) HCN and CO on Neptune: An intrinsic origin, *Bull. Am. Astron. Soc.* **24**, 972, and Paper presented at the Division of Planetary Sciences Meeting, AAS, Munich, Germany, October, 1992.
5. Gautier, D., Conrath, B.J., Owen, T., De Pater, I and Atreya, S.K. (1994) The Troposphere of Neptune, D. Cruikshank (ed.), *Neptune*, University of Arizona Press, Tucson, in press.
6. Bishop, J., Atreya, S.K., Romani, P.N., Orton, G.S., Sandel, B.R. and Yelle, R.V. (1994) The Middle and Upper Atmosphere of Neptune, D. Cruikshank (ed.), *Neptune*, University of Arizona Press, Tucson, in press.
7. Lellouch, E., Romani, P.N. and Rosenqvist, J. (1994) The vertical distribution and origin of HCN in Neptune's atmosphere, *Icarus* **108**, 112-136.
8. Conrath, B.J., Gautier, D., Owen, T.C. and Samuelson, R.E. (1993) Constraints on N_2 in Neptune's atmosphere from Voyager measurements, *Icarus* **101**, 168-172.
9. Moses, J.I., Allen, M. and Yung, Y.L. (1992) Hydrocarbon nucleation and aerosol formation in Neptune's atmosphere, *Icarus* **99**, 318-346.
10. Schopper, H. (1973) Elastic and charge exchange scattering of elementary particles, in H. Helwege (ed.), *Landolt-Bornstein Numerical Data and Functional Relationships in Science and Technology*, Springer-Verlag, New York, pp.7-30.
11. Belletini, G., Cocconi, G., Diddens, A.N., Lillethun, E., Matthide, G., Scanlon, J.P. and Wetherell, A.M.

(1966) Proton-nuclei cross sections at 20 GeV, *Nucl. Phys.* **29**, 609-624.

12.   Rapp, D., Englander-Golden, P. and Briglia, D.D. (1965) Cross sections for dissociative ionization of molecules by electron impact, *J. Chem. Phys.* **42**, 4081-4085.

13.   Stevens, M.H., Strobel, D.F., Summers, M.E. and Yelle, R.V. (1992) On the thermal structure of Triton's thermosphere, *Geophys. Res. Lett.* **19**, 669-672.

14.   Selesnick, R.S. and Stone, E.C. (1991) Neptune's cosmic ray cutoff, *Geophys. Res. Lett.* **18**, 361-364.

15.   Bar-Nun, A., Kleinfeld, I. and Kochavi, E. (1988) Trapping of gas mixtures by amorphous water ice, *Phys. Rev. B* **38**, 7749-7754.

16.   Owen, T.C. and Bar-Nun, A. (1994) Comets, Impacts, and Atmospheres, *Icarus*, submitted.

THE CHEMICAL ATMOSPHERIC COMPOSITION OF THE GIANT PLANETS

T. ENCRENAZ
DESPA, Observatoire de Paris
F-92195 Meudon

Abstract

For a long time it was believed that the atmospheres of the giant planets, dominated by molecular hydrogen and helium, were similar in composition to the primordial nebula from which they formed. However, this image has strongly evolved over the past twenty years, due to new developments of ground-based infrared spectroscopy, coupled with the success of the Voyager space mission.

Significant differences were measured in the abundances of helium, deuterium and carbon of the four giant planets. The variations in the C/H and D/H ratios have given support to the "nucleation" formation scenario, in which the four giant planets first accreted a nucleus of about ten terrestrial masses, big enough to bind gravitationally the surrounding gaseous nebula; the helium depletion in Saturn has been interpreted as a differentiation effect in Saturn's interior; the apparent helium excess in Neptune, coupled with the recent unexpected detection of CO and HCN in this planet, might imply the presence of molecular nitrogen. In the case of Jupiter and Saturn, disequilibrium species have been detected (CO, PH_3, GeH_4, AsH_3), which are tracers of vertical dynamical motions.

In the future, significant progress in our knowledge of the Jovian composition, including the noble gases, should be obtained with the mass spectrometer of the Galileo probe. The ISO mission is expected to provide new far-infrared spectroscopic data which should lead to the detection of new minor species and a better determination of the D/H ratio.

1. Introduction

It has been known for many decades that the atmospheres of the giant planets were mostly composed of hydrogen and helium, with traces of minor compounds in a reduced form. Methane and ammonia were first detected from their near-infrared spectral signatures (Wildt, 1932); hydrogen, expected to be the dominant constituent (Herzberg, 1952) was detected almost thirty years later (Kiess *et al.*,1960). Helium, expected to be present at the 10% level, was indirectly inferred from the determination of the mean molecular weight, measured by a stellar occultation experiment (Baum and Code, 1953).

Earth, Moon, and Planets 67: 77–87, 1995.

The hydrogen-rich composition of the giant planets was understood as an effect of their very high escape velocities. A molecule can escape an atmosphere if its kinetic velocity exceeds its escape velocity. This is more likely if the temperature is high and if the planet is small; in the case of the giant planets, the escape velocity is so high that even the lightest element, hydrogen, cannot escape over the lifetime of the Solar system (Spitzer, 1952). Since the giant planets were massive enough to keep all their constituents by gravitation, their atmospheres were thus expected to reflect closely the chemical composition of the primordial nebula from which they formed. Since abundances were assumed to be solar (or "cosmic") in the gaseous phase of the primordial nebula (with the exception of deuterium, continuously destroyed in stars), abundance ratios in the giant planets were more or less expected to follow the cosmic values.

From the early sixties to the mid-seventies, there was little progress in our knowledge of the atmospheric composition of the giant planets. Then, within a few years, a large number of minor species were detected (Table 1), due to the fast development of ground-based infrared techniques (see Atreya, 1986 and Encrenaz, 1990 for a review). The infrared spectral range is indeed best suited for studying the vibration-rotation bands of neutral molecules; in addition, these cold objects radiate a large fraction of their thermal energy in the infrared, beyond 5 μm. Infrared spectrosocopy at various wavelengths allowed astronomers to probe different altitude levels of the atmospheres; in particular, the 5-μm window, observable from the ground and free from methane and ammonia absorption, was very important for detecting minor tropospheric species on Jupiter (Ridgway et al., 1976) and later Saturn (Prinn et al., 1984). This method allowed the unexpected identification of disequilibrium species like PH_3, GeH_4, AsH_3 and CO in the tropospheres of Jupiter and Saturn.

Another major step was the Voyager mission with its two spacecraft which encountered Jupiter in 1979 and Saturn in 1980 and 1981; Voyager 2 encountered Uranus in 1986 and Neptune in 1989. Their infrared interferometers, named "IRIS", measured the infrared spectra of the four giant planets and Titan in the 5-50 μm range (200-2000 cm[-1]), with a spectral resolution of 4.3 cm[-1] (Hanel et al., 1979a&b, 1981, 1982, 1986; Conrath et al., 1990). These data, together with the radio-occultation Voyager measurements, allowed a simultaneous retrieval of the helium abundance and the vertical distribution of the thermal profiles on the four giant planets. Determinations of the C/H and D/H ratios on Jupiter and Saturn were also obtained from the study of the emission bands of CH_4 at 7.7 μm and CH_3D at 8.6 μm.

From all these results, a new picture has emerged, showing evidence for significant differences between the chemical composition of the giant planets (Table 1). An enrichment is observed, for both carbon and deuterium, from Jupiter to Neptune; the helium abundance strongly varies, with a minimum value in the case of Saturn and a maximum value in the case of Neptune. As a last example of these differences, CO and HCN have been detected recently in the stratosphere of Neptune (Marten et al., 1991,1993; Rosenqvist et al., 1992), but not in the stratospheres of the other giant planets; their abundances in Neptune are significantly higher than the predicted values.

These important results have been used to revisit the formation and evolution models of the giant planets. It now appears that the formation scenarios of the giant planets were likely to be much more complex than previously thought. By gathering all information coming from objects formed in this range of heliocentric distances (5 to 50 AU), including giant planets, satellites, Pluto and the comets, one should be able to proceed one step further in our knowledge of early formation and evolution processes in the outer solar system.

TABLE 1. Gaseous molecular species in the giant planets (summarized and updated from Atreya, 1986)

SPECIES	JUPITER	SATURN	URANUS	NEPTUNE
		MIXING RATIOS (VERSUS H_2)		
H_2	1	1	1	1
HD	6×10^{-5}	1×10^{-4}	***	***
He	0.11	0.03	0.18	0.23
CH_4	2×10^{-3}	4×10^{-3}	2×10^{-2}	4×10^{-2}
$^{13}CH_4$	2×10^{-5}	4×10^{-5}		
CH_3D	3×10^{-7}	4×10^{-7}	1×10^{-5}	3×10^{-5}
NH_3	3×10^{-4}	2×10^{-4}		
$^{15}NH_3$	1×10^{-6}			
C_2H_2	3×10^{-8} (*)	7×10^{-8} (*)	1×10^{-7}	$1\text{-}9 \times 10^{-7}$
$^{12}C^{13}CH_2$	3×10^{-9} (*)			
C_2H_6	2×10^{-6}	3×10^{-6}		3×10^{-6}
C_3H_4	3×10^{-9} (**)			
C_3H_8	6×10^{-7} (**)			
C_2H_4	7×10^{-9} (**)			
C_6H_6	2×10^{-9} (**)			
H_2O	1×10^{-6} (*)			
CO	1.5×10^{-9} (+)	2×10^{-9} (+)		6×10^{-7} (++)
PH_3	5×10^{-7}	1.5×10^{-6}		
GeH_4	7×10^{-10}	4×10^{-10}		
AsH_3	$3 \times 10\text{-}10$	2.4×10^{-9}		
HCN	2×10^{-9}			3×10^{-10} (++)
H_3^+	(++)	(++)		

NOTES: Mixing ratios only indicate orders of magnitude
(*)   variable
(**)   in the north polar region
(***)   detected; mixing ratio is very uncertain
(+)   detected in the troposphere
(++)   detected in the stratosphere

2. Constraints in formation and evolution processes

The determination of elemental and isotopic ratios in the giant planets has provided significant constraints upon the formation and evolution of these objects. These ratios (D/H, C/H, N/H...) have been derived from the study of atmospheric species (HD, CH_4, NH_3), and the determination of their abundances and vertical distributions. The measurement of these quantities implies the simultaneous knowledge of the thermal profile T(P), which, in the case of the four giant planets, has been retrieved with good accuracy from the Voyager radio-occultation and far-infrared experiments. Once the thermal profile is known, vertical distributions of atmospheric species can be deduced from the measurement of their infrared spectroscopic bands, and a comparison made with synthetic models.

2.1 THE CARBON AND DEUTERIUM ABUNDANCES

Carbon was the first species for which an abundance variation was suspected among the four giant planets. The C/H ratio is derived from the CH_4/H_2 ratio, which, in the case of Jupiter and Saturn, is believed to be constant over the whole atmospheric range probed in the infrared. On Uranus and Neptune, saturation is expected to take place in the region of the upper troposphere and above, so that the CH_4 measurement, in order to be meaningful in terms of C/H determination, has to be performed in the deep troposphere. The first indication for a variation in CH_4 came from the analysis of the visible and near-infrared methane bands, which suggested a strong CH_4 enhancement on Uranus and Neptune, with respect to the two others (Encrenaz et al.,1974; Lutz et al., 1976). In the case of Jupiter and Saturn, an accurate determination of C/H was obtained from the analysis of the 7.7 μm CH_4 band observed with the Voyager IRIS experiment (Gautier et al., 1982; Courtin et al., 1984). These data, coupled with other analyses in the near-IR range (Combes and Encrenaz, 1979; Buriez and de Bergh, 1981) provided definitive evidence for a carbon enrichment with respect to the solar value, by a factor of 2 in the case of Jupiter and around 4 (+/-2) in the case of Saturn. In the case of Uranus and Neptune the carbon enrichment is believed to be at least 10 times larger than for Jupiter; present analyses suggest a carbon enrichment in the range 25 to 60 (Baines and Bergstrahl, 1986; Baines and Smith, 1990; Fegley et al., 1991; Gautier et al., 1994).

The determination of a carbon excess which significantly increases from Jupiter to Neptune has given decisive support to the "nucleation" formation theory of the giant planets. Two decades ago, a "homogeneous" model was suggested, assuming giant homogeneous gaseous protoplanets with chemical compositions close to solar abundances. However, this model cannot account for the carbon abundance variations observed over the past ten years. In contrast, the nucleation models developed by Perri and Cameron (1974) and Mizuno (1980) does predict an enrichment in heavy elements. In this model, cores of heavy elements are formed first, in comparable masses for all giant planets (about 10 terrestrial masses). This mass is sufficient for the surrounding gaseous nebula to collapse. In the core formation, accretional heating leads to outgassing from ices (CO, CH_4, NH_3, H_2O...) and/or clathrates/hydrates, and a subsequent enrichment of the outer envelope in heavy elements like C, O, N, etc. As pointed out by Stevenson (1982), infalling planetesimals might also have contributed to the observed enrichment. Since the relative size of the core grows from Jupiter to Neptune, the C/H variations over the four planets are in qualitative agreement with this model (Gautier and Owen, 1989).

If the nucleation model is valid, all elements heavier than H, He and Ne (which are not expected to be trapped in ices) are expected to be enhanced in the atmospheres of the giant planets. We would thus expect to observe excesses of NH_3, H_2O, etc. However these measurements are very difficult, because these molecules condense in the four giant planets to form clouds, and a measurement below the clouds is very difficult by remote sensing spectroscopy. NH_3 seems to be enriched on Jupiter and Saturn (de Pater, 1986), but depleted on Uranus and Neptune (de Pater and Mitchell, 1993), which might be due to cloud chemistry, but is not completely understood at the present time (Gautier et al.,1994). H_2O has been detected in Jupiter only (Larson et al., 1975); Carlson et al., 1992); other sources of oxygen may exist, as illustrated by the detection of CO (Beer, 1975). Methane, which does not condense in Jupiter and Saturn, is considered as the best test for measuring the excess on heavy elements in the giant planets.

Deuterium is another isotope of particular interest for bringing constraints on formation scenarios. Unlike the other elements, the deuterium abundance in the protosolar nebula is higher than in the Sun, where deuterium is converted into helium. The protosolar value is about 3×10^{-5}, as derived from ^3He solar wind measurements (Geiss, 1993). It is also expected to be higher than its value in the local interstellar medium today, because of the continuous destruction of deuterium in stars since the formation of the Solar system, 4.5 billion years ago. This variation of the D/H ratio as a function of time can actually be estimated from a modelling of the chemical evolution of galaxies (Delbourgo-Salvador et al., 1985). The observed mean value of about 10^{-5} in the local interstellar medium (Linsky et al., 1993) is consistent with these models.

Deuterium has been detected in the four giant planets, from both the HD and CH_3D molecules. The HD detection (Trauger et al.,1973; Macy and Smith, 1978) directly leads to D/H (the HD mixing ratio being twice the D/H ratio), but the HD lines appear in the visible region and are affected by scattering and blends with weak methane lines. In the case of Jupiter and Saturn, infrared bands of CH_3D (Beer et al., 1972, Fink and Larson, 1978) provide a less model-dependent estimate of the CH_3D abundance, but its conversion into D/H is a function of the fractionation factor f through the equation D/H $= 1/4f \times (CH_3D/CH_4)$; f is a strong function of temperature (Fegley and Prinn, 1988). For Uranus and Neptune, near-infrared bands of CH_3D are used (de Bergh et al., 1984, 1990). As a result, in spite of large error bars, the D/H in Jupiter and Saturn seems to be close to the primordial value of 3×10^{-5}, while the Uranus and Neptune values are a factor of 5 to 10 larger (Gautier and Owen, 1989; Owen, 1992).

In the homogeneous formation model of the giant planets, one would expect the deuterium abundance in all giant planets to be similar to the primordial nebula value. In the nucleation model, in contrast, the deuterium coming from the core might be significantly enriched, as observed in molecules trapped in ices in other Solar system objects, like Titan (Owen et al., 1986) and comet Halley (Eberhardt et al., 1987). As first suggested by Owen et al. (1986), the origin of the D/H enrichment in this second deuterium reservoir could be ion-molecule reactions, as observed in the interstellar medium, which would have occurred before the formation of planetesimals. As discussed by Gautier and Owen (1989), the observed excess of D/H in Uranus and Neptune would be in qualitative agreement with both the nucleation model and the "ice reservoir" of deuterium proposed by Owen et al. (1986); however, more accurate data are needed before any firm conclusion can be drawn.

2.2 THE HELIUM ABUNDANCE

In the Big Bang theory, helium is assumed to have formed mostly in the first stage of the Universe. If there was no differentiation effect within the giant planets, a measurement of H_2/He in these objects would give a direct measurement of the protosolar helium abundance.

The He I resonance line at 584 A , detected on Jupiter, Saturn and Uranus from the Voyager spacecraft (Broadfoot et al., 1981; Sandel et al., 1982, Broadfoot et al., 1986), is formed above the homopause and cannot be used for a determination of the helium relative abundance. The H_2/He ratio has been indirectly derived from a combination of two methods: the inversion of the far-infrared spectra of the giant planets, as measured with the IRIS Voyager experiment, and the determination of the mean molecular weight derived from the Voyager radio-occultation experiments. Large variations were observed in the helium mass fraction: the derived values were (assuming pure H_2-He atmospheres) 0.18 for Jupiter (Gautier et al., 1981), 0.06 for Saturn (Conrath et al., 1984), 0.26 for Uranus (Conrath et al., 1987) and 0.32 for Neptune (Conrath et al., 1991); the estimated value for the proto-Sun would be 0.28 (Sackman et al., 1990). The helium depletion in Saturn was interpreted as a differentiation effect in Saturn's interior, due to the condensation, in metallic hydrogen, of helium droplets sinking toward the center and depleting the helium content of the outer envelope; the same effect might take place on Jupiter and explain its helium deficiency with respect to the primordial value. It would be less efficient than in the case of Saturn, as helium condensation is expected to have occurred later in the cooling process of both planets, Jupiter starting from a higher temperature. In the case of Uranus and Neptune, the pressure of their interiors does not seem sufficient for metallic hydrogen to be present, so that the helium condensation effect is not expected to take place (Gautier and Owen, 1989). Indeed, the Uranus value appears to be representative of the primordial helium abundance. The helium abundance of Neptune is marginally compatible with the primordial value. However, the value derived for a pure H_2-He atmosphere shows a slight excess with regard to this value. A possible explanation might be that molecular nitrogen is present in its interior at the 0.3% level, while the helium abundance would be, as in Uranus, primordial (Conrath et al., 1993). This nitrogen abundance would explain the molecular weight excess possibly detected on Neptune, and could be responsible for the presence of stratospheric HCN in this planet (Gautier et al.,1994). This result, if confirmed, would have important implications upon the chemical nature of planetesimals from which the giant planets formed. Finally, an important question which remains to be solved is the origin of the difference between Uranus and Neptune: why are HCN and CO present in the stratosphere of Neptune, but not of Uranus? why would nitrogen be present in the troposphere of Neptune, but not of Uranus? As suggested by Conrath et al. (1993) and Hubbard et al. (1994), the origin might come from the difference in the internal energy sources of the two planets. In the case of Uranus, the internal heat flux is so small that convection might be inhibited in several layers of the troposphere, preventing upward movement of N_2 from the deep interior.

3. Spatio-temporal variations

Several processes can presently alter the chemical compositions of the giant planets, leading to non-uniform vertical mixing and spatio-temporal variations. These processes include condensation, photochemistry, interaction with the magnetosphere and general circulation.

As an example, the study of vertical motions in the giant planets can significantly benefit from the analysis of non-equilibrium species. These minor atmospheric compounds should not be observable on the basis of chemical equilibrium models, but are carried into visibility by upward vertical motions on timescales shorter than their chemical lifetime (Drossart *et al.*, 1989). The most abundant species of this kind is phosphine, which is present in solar abundance in Jupiter's deep troposphere (Ridgway *et al.*,1976), and is enriched by a factor about 3 in the case of Saturn (Bregman *et al.*, 1975). Other disequilibrium species include CO, GeH_4 and AsH_3, all detected at deep atmospheric levels in the 5 μm window (Beer, 1975; Noll *et al.*, 1986; Bézard *et al.*,1989; Noll *et al.*,1989). In the same way, spatio-temporal variations of condensable compounds such as H_2O and NH_3 can also be used for deriving information about vertical atmospheric motions (Lellouch *et al.*,1989).

The existence of aurorae on Jupiter and Saturn has been known for a long time from the Ly alpha emission of these planets (Clarke *et al.*, 1980, 1989; Broadfoot *et al.*, 1981). These regions were also characterized by a strong infrared emission in the bands of methane and various hydrocarbons (Caldwell *et al.*, 1980; Kim *et al.*,1985; Drossart *et al.*, 1986). This study has known a renewed interest with the detection of the H_3^+ ion in Jupiter's auroral regions (Drossart *et al.*,1989, 1992; Oka and Geballe, 1990), in the near-infrared range. These emissions appear to be powerful tracers of the thermal profile in the upper stratosphere (Drossart *et al.*, 1993).

4. Open problems and future studies

Our understanding of the giant planets has been completely renewed over the past decades. Following the exploration of the four planets by the Voyager spacecraft, and with the support of a very active ground-based observation program, we have successively discovered more and more differences, which are all indicators of their past or present history. In the light of these new results, there are still many important questions which remain to be solved.

One of the basic issues is a precise determination of C/H, D/H and H/He. Unfortunately, an improvement of the C/H determination on Uranus and Neptune seems difficult, because the thermal infrared range gives only access to the upper troposphere and the stratosphere, where methane is likely to condense. The case of deuterium, in contrast, is more favorable. A reliable determination of D/H on the four giant planets should be achievable through the HD rotational transitions, in particular with the spectrometers of the ISO satellite, to be launched in 1995 (Encrenaz and Kessler, 1992). These instruments also should provide an improved determination of the H_2/He ratio. A measurement of the noble gas abundances would be crucial for assessing new constraints upon formation scenarios; hopefully this will be performed in the case of Jupiter by the Galileo probe.

Global monitoring of the planets is necessary to understand the photochemical processes now governing their chemical compositions (photochemistry, general circulation). In addition to the exploration of Jupiter and Saturn by Galileo and Cassini respectively, one can foresee promising improvements in the field of ground-based imaging spectroscopy. Present developments include, in particular, infrared cameras with improved spatial resolution and sensitivity. The combination of a Fourier-Transform spectrometer with a bidimensional infrared camera will allow, in particular, a complete mapping of Jupiter and Saturn in the 5- μm window, and thus a continuous monitoring of their tropospheric compositions; the first adaptive optics instruments are now being developed (Saint-Pé *et al.*, 1993a&b), and will provide an improvement by a factor of 10 in the spatial resolution of the infrared images.

Finally, the spectrographs of the ISO mission could allow the detection of new minor species, especially in the far-infrared range. This search will also be possible from the ground, with the development of submillimeter interferometers and heterodyne receivers.

5.   References

Atreya, S.K. (1986) *Atmospheres and ionospheres of the outer planets and their satellites*, Springer-Verlag.

Baines, K.H. and Bergstralh, J.T. (1986) The structure of the uranian atmosphere: Constraints from the geometric albedo spectrum and H_2 and CH_4 line profiles, *Icarus* 65, 406-441.

Baines, K.H. and Smith, W.H. (1990) The atmospheric structure and dynamical properties of Neptune derived from ground-based and IUE spectrophotometry, *Icarus* 85, 65-108.

Baum, W.A. and Code, A.D. (1953) A photometric observation of the occultation of s Arietis by Jupiter, *Astron. J.* 58, 108-112.

Beer, R. (1975) Detection of carbon monoxide in Jupiter, *Astrophys.J.* 200, L167-L169.

Beer, R., Farmer, C.B., Norton, R.H., Martonchik, J.V., and Barnes, T.G. (1972) Observation of deuterated methane in the atmosphere, *Science* 175, 1360-1361.

Bézard, B., Drossart, P., Lellouch, E., Tarrago, G. and Maillard, J.P. (1989) Detection of Arsine in Saturn, *Astrophys.J.* 346, 509-513.

Bregman, J.D., Lester, D.F., and Rank, D.M. (1975) Observation of the v_2 band of PH_3 in the atmosphere of Saturn, *Astrophys. J.* 202, L55-L56.

Broadfoot, A.L. *et al.* (1981) Overview of the Voyager ultraviolet spectrometry results through Jupiter encounter, *J. Geophys. Res.* 86, 8259-8284.

Broadfoot, A.L. *et al.* (1986) Ultraviolet spectrometer observations of Uranus, *Science* 233, 74-79.

Buriez, J.C. and de Bergh, C. (1981) A study of the atmosphere of Saturn based on methane line profiles near 1.1 μm, *Astron. Astrophys.* 94, 382-390.

Caldwell, J., Tokunaga, A.T. and Gillett, F.C. (1980) Possible infrared aurorae on Jupiter, *Icarus* 44, 666-675.

Carlson, B.E., Lacis, A.A. and Rossow, W.B. (1992) The abundance and distribution of water vapor in the Jovian troposphere as inferred from Voyager IRIS observations, *Astrophys.J.* 388, 648-668.

Clarke, J.T., Moos, H.W., Atreya, S.K. and Lane, A.L. (1980) Observations from earth orbit and variability of the polar aurora on Jupiter, *Astrophys. J.* **241**, L179-L182.

Combes, M. and Encrenaz, T. (1979) A method for the determination of abundance ratios in the outer planets - Application to Jupiter, *Icarus* **39**, 1-27.

Conrath, B., Gautier, D., Hanel, R. and Hornstein, J.S. (1984) The helium abundance of Saturn from Voyager measurements, *Astrophys. J.* **282**, 807-815.

Conrath, B.J., Gautier, D., Hanel, R., Lindal, G. and Marten, A. (1987) The helium abundance of Uranus from Voyager measurements, *J. Geophys. Res.* **92**, 15003-15010.

Conrath, B.J., Gautier, D., Lindal, G., Samuelson, R.E. and Shaffer, W.A. (1991) The helium abundance of Neptune from Voyager measurements, *J. Geophys. Res.* **96**, 18907-18919.

Conrath, B.J., Gautier, D., Owen, T. and Samuelson, R.E. (1993) Constraints on N_2 in Neptune's atmosphere from Voyager measurements, *Icarus* **101**, 168-171.

Conrath, B. *et al.* (1989) Infrared observations of the Neptunian system, *Science* **246**, 1454-1459.

Courtin, R., Gautier, D., Marten, A., Bézard, B. and Hanel, R. (1984) The composition of Saturn's atmosphere at northern temperate latitudes from Voyager IRIS spectra : NH_3, PH_3, C_2H_2, C_2H_6, CH_3D, CH_4 and the Saturnian D/H isotopic ratio, *Astrophys.J.* **287**, 899-916.

Delbourgo-Salvador, P., Gry, C., Malinie, G. and Audouze, J. (1985) Effects of nuclear uncertainties and chemical evolution of the standard big bang nucleosynthesis, *Astron. Astrophys.* **150**, 53-61.

de Bergh, C., Lutz, B., Owen, T. and Chauville, J. (1988) Monodeuterated methane in the outer solar system. II. Its detection on Uranus at 1.6 μm, *Astrophys. J.* **311**, 501-510.

de Bergh, C., Lutz, B., Owen, T. and Maillard, J.P. (1990) Monodeuterated methane in the outer solar system. IV. Its detection and abundance on Neptune, *Astrophys. J.* **355**, 661-666.

de Pater, I. (1986) Jupiter's Zone-Belt structure at radio-wavelengths. II. A comparison of observations with model-atmosphere calculations, *Icarus* **68**, 344-365.

de Pater, I. and Mitchell, D.L. (1983), Microwave observations of the planets : The importance of laboratory measurements, *J. Geophys. Res. Planets* **98**, 5471-5490.

Drossart, P., Bézard, B., Atreya, S.K., Lacy, J., Serabyn, E., Tokunaga, A.T. and Encrenaz, T. (1986) Enhanced acetylene emission near the north pole of Jupiter, *Icarus* **66**, 610-618.

Drossart, P., Courtin, R., Atreya, S. and Tokunaga, A. T. (1989) Variations in the Jovian atmospheric composition and chemistry, in M.Belton, R.West and J.Rahe (edts), *Time-variable phenomena in the Jovian system*, NASA SP-494, pp.344-362.

Drossart, P., Maillard, J.P., Caldwell, J., Kim, S.J., Clarke, J., Waite, H., and Wagener, R. (1989) Detection of H_3^+ in Jupiter, *Nature* **340**, 539-541.

Drossart, P., Prangé, R. and Maillard, J.P. (1992) Morphology of infrared H_3^+ emissions in the auroral regions of Jupiter, *Icarus* **97**, 10-25.

Drossart, P., Maillard, J.P., Caldwell, J. and Rosenqvist, J. (1993) Line-resolved spectroscopy of the Jovian H_3^+ auroral emission at 3.5 micrometers, *Astrophys.J.* **402**, L25-L28.

Eberhardt, P., Dolder, U., Schulte, W., Krankowsky, D., Lämmerzahl, P., Hoffman, J.H., Hodges, R.R., Berthellier, J.J. and Illiano, J.M. (1987) The D/H ratio in water from comet P/Halley, *Astron. Astrophys.* **187**, 435-437.

Encrenaz, T. (1990) Remote sensing of the atmospheres of Jupiter, Saturn and Titan. *Rep. Prog. Phys.* **53**, 793-836.

Encrenaz, T. and Kessler, M.F. (1992) *Infrared Astronomy with ISO*, Nova Science.

Encrenaz, T., Hardorp, J., Owen, T. and Woodman, J. H. (1974) Observational constraints on model atmospheres for Uranus and Neptune. In A.Woscyk and I. Iwaniszewska (edts) *Exploration of the Planetary Systems*, Dortrecht : D.Reidel, pp.487-496.

Fegley, M.B. and Prinn, R.G. (1988) The predicted abundances of deuterium-bearing gases in the atmospheres of Jupiter and Saturn, *Astrophys.J.* **326**, 490-508.

Fegley, B., Gautier, D, Owen, T. and Prinn, R.G. (1991) Spectroscopy and chemistry of the atmosphere of Uranus, in *Uranus* J.T.Bergstralh *et al.* edts, Univ. of Arizona, 147-203.

Fink, U. and Larson, H.P. (1978) Deuterated methane observed on Saturn, *Science* **201**, 343-345.

Gautier, D., Bézard, B., Marten, A., Baluteau, J. P., Scott, N., Chédin, A., Kunde, V. and Hanel, R. (1982) The C/H ratio in Jupiter from the Voyager IRIS investigation, *Astrophys.J.* **257**, 901-912.

Gautier D and Owen T (1989) The composition of outer planet atmospheres, in S. Atreya, J. Pollack and M. Matthews (edts) *Origin and evolution of planetary and satellite atmospheres*, pp.487-512.

Gautier, D., Conrath, B., Flasar, M., Hanel, R., Kunde, V., Chedin, A. and Scott,N. (1981) The helium abundance of Jupiter from Voyager, *J. Geophys. Res.* **86**, 8713-8720.

Gautier, D., Conrath, B. J., Owen, T., de Pater, I. and Atreya, S. K. (1994), *The troposphere of Neptune*, in *Neptune*, ed. D.Cruikshank *et al.*, The University of Arizona Press, Tucson (in press).

Geiss, J. (1993) Primordial abundances of hydrogen and helium isotopes, in N. Prantzos *et al.* (edts) *Origin and evolution of the elements*, Cambridge, pp. 89-106.

Hanel, R. *et al.* (1979a) Infrared observations of the Jovian system from Voyager 1, *Science* **204**, 972-976.

Hanel, R. *et al.* (1979b) Infrared observations of the Jovian system from Voyager 2, *Science* **206**, 952-956.

Hanel, R. *et al.* (1981) Infrared observations of the Saturnian system from Voyager 1, *Science* **212**, 192-200.

Hanel, R. *et al.* (1982) Infrared observations of the Saturnian system from Voyager 2, *Science* **215**, 544-548.

Hanel,R. *et al.*, Infrared observations of the Uranus sytem, *Science* **233**, 70-74.

Herzberg, G. (1952) in G.P. Kuiper (edt) *The Atmospheres of the Earth and Planets*, Chicago.

Hubbard, W.B., Pearl, J.C., Podolak, M. and Stevenson, D.J. (1994), *The interior of Neptune*, in *Neptune*, D.Cruikshank *et al.* (edt), The Univ. of Arizona Press (in press).

Kiess, C.C., Korliss, C.H. and Kiess, H.K. (1960) High-dispersion spectra of Jupiter, *Astrophys. J.* **132**, 221-231.

Kim, S.J., Caldwell, J., Rivolo, A.R., Wagener, R. and Orton, G.S. (1985) Infrared Polar Brightening of Jupiter : III- Spectrometry from the Voyager 1 IRIS Experiment, *Icarus* **64**, 233-248.

Larson, H.P., Fink, U., Treffers, R. and Gautier, T.N. (1975) Detection of water vapor on Jupiter, *Astrophys. J.* **197**, L-137-L140.

Lellouch, E., Drossart, P. and Encrenaz, T. (1989) A new analysis of the Jovian 5-μm Voyager IRIS spectra, *Icarus* **77**, 457-465.

Linsky, J.L., Brown, A., Gaylay, K., Diplas, A., Savage, B.D., Ayres, T.R., Landsman, W., Shore, S.N. and Heap, S.R. (1993) Goddard high-resolution spectrograph observations of the local interstellar medium and the deuterium/hydrogen ratio along the line of sight toward Capella, *Astrophys. J.* **402**, 694-709.

Lutz, B. L., Owen ,T. and Cess, R. D. (1976) Laboratory band strengths of methane and their application to the atmospheres of Jupiter, Saturn, Uranus, Neptune and Titan, *Astrophys. J.* **203**, 541-551.

Macy, W.W. and Smith, W.H. (1978), Detection of HD on Saturn and Uranus and the D/H ratio, *Astrophys. J.* **222**, L137-L140.

Marten , A., Gautier, D., Owen, T., Sanders, D., Tilanus, R.T., Deane, J. and Matthews, H. (1991) First detections of CO and HCN in the atmosphere of Neptune, *B.A.A.S* **23**, 1164.

Marten, A., Gautier, D., Owen,T., Sanders, D.B., Matthews, H.E., Atreya, S.K., Tilanus, R.P.J., and Deane, J.R. (1993) First observations of CO and HCN on Neptune and Uranus at millimeter wavelengths and their implications for atmospheric chemistry, *Astrophys. J.* **406**, 285-297.

Mizuno, H. (1980) Formation of the giant planets, *Prog. Theor. Phys.* **64**, 544-557.

Noll, K.S., Knacke, R.F., Geballe, T.R. and Tokunaga, A.T. (1986) Detection of carbon monoxide in Saturn, *Astrophys. J.* **309**, L91-L94.

Noll, K.S., Geballe, T.R. and Knacke, R.F. (1989) Arsine in Saturn and Jupiter, *Astrophys. J.* **338**, L71-L74.

Oka, T. and Geballe, R.T. (1990) Observation of the 4 μm fundamental band of H_3^+ in Jupiter, *Astrophys. J.* **351**, L53-L56.

Owen, T (1992), Deuterium in the Solar System, in P. Singh (edt.) *Astrochemistry of cosmic phenomena*, Dordrecht-Kluwer, pp. 97-101.

Owen, T., Lutz, B. L. and de Bergh, C. (1986) Deuterium in the outer solar system: Evidence for two distincts reservoirs, *Nature* **320**, 244-246.

Perri, F. and Cameron, A.G.W. (1974) Hydrodynamic instability of the solar nebula in the presence of a planetary core, *Icarus* **22**, 416-425.

Prinn, R.G., Larson, H.P., Caldwell, J.J and Gautier, D. (1984), *Composition and Chemistry of Saturn's Atmosphere*, in *Saturn*, T. Gehrels and M.S. Matthews (edts.), Univ. of Arizona, 88-149

Ridgway, S.T., Larson, H.L. and Fink, U. (1976), *The Infrared Spectrum of Jupiter*, in "Jupiter", T. Gehrels (edt.), Univ. of Arizona, 384-417

Rosenqvist, J., Lellouch, E., Romani, P., Paubert, G. and Encrenaz, T. (1992) Millimeter-wave observations of Saturn, Uranus et Neptune : CO and HCN on Neptune, *Astrophys. J.* **392**, L99-L102.

Sackman, I.J., Boothroyd, A.I. and Fowler, W.A. (1990) Our Sun. I. The standard model : successes and failures, *Astrophys. J.* **360**, 727-736.

Sandel, B.R. *et al.* (1982) Extreme ultraviolet observations from the Voyager 2 encounter with Saturn, *Science* **215**, 548-553.

Saint-Pé, O., Combes, M., Rigaut, F., Tomasko, M. and Fulchignoni, M. (1993a) Demonstration of adaptive optics for resolved imagery of Solar-system objects : Preliminary results on Pallas and Titan, *Icarus* **105**, 263-270.

Saint-Pé, O., Combes, M. and Rigaut, F. (1993b) Ceres surface properties by high-resolution imaging from Earth, *Icarus* **105**, 271-281.

Spitzer, L. (1952), in G.P. Kuiper (edt.), *The Atmospheres of the Earth and Planets*, Chicago.

Stevenson, D.J. (1982), Formation of the giant planets, *Planet. Space Sci.* **30**, 755-764.

Trauger, J., Roesler, F., Carleton N,P, and Traub, W.A. (1973) Observation of HD on Jupiter and the D/H ratio, *Astrophys. J.* **184**, L137-L141.

Wildt, R. (1932) Veröff Univ. Sternwarte Göttingen 2 22 171

NH₃, H₂S, AND THE RADIO BRIGHTNESS TEMPERATURE SPECTRA OF THE GIANT PLANETS

THOMAS R. SPILKER

Jet Propulsion Laboratory / California Institute of Technology
Pasadena, CA, USA 91109

Recent radio interferometer observations of Neptune enable comparisons of the radio brightness temperature (T_B) spectra of all four giant planets. This comparison reveals evidence for fundamental differences in the compositions of Uranus' and Neptune's upper tropospheres, particularly in their ammonia (NH_3) and hydrogen sulfide (H_2S) mixing ratios, despite those planets' outward similarities. The tropospheric abundances of these constituents yield information about their deep abundances, and ultimately about the formation of the planets from the presolar nebula (Atreya *et al.*, 1995).

Figures 1, 2, 3, and 4 show the T_B spectra of Jupiter, Saturn, Uranus, and Neptune, respectively, from 0.1 to tens of cm wavelength. The data shown are collected from many observers. Data for Jupiter, Saturn, and Uranus are those cataloged by de Pater and Massie (1985), plus the Saturn Very Large Array (VLA) data by Grossman *et al.* (1989). Figure 3, Uranus, shows only data acquired since 1973. Before 1973 Uranus' T_B increased steadily as its pole moved into view, causing significant scatter in those data. Neptune data at >1 cm, all taken at the VLA, are collected from de Pater and Richmond (1989), de Pater *et al.* (1991), and Hofstadter (1993). For a variety of reasons, such as susceptibility to source confusion, single-dish data at those wavelengths are much noisier than the more reliable VLA data and have been ignored. Single-dish data by Griffin and Orton (1993) shortward of 0.4 cm are shown, along with the Owens Valley Radio Observatory (interferometer) datum at 0.266 cm by Muhleman and Berge (1991).

Spectra of Jupiter, Saturn, and Neptune share certain gross characteristics. In each spectrum, T_B at 1.3 cm is ~120-140 K, less than ~30 K different from that at 0.1 cm. All three spectra show a break in slope at or near 1.3 cm, with T_B increasing fairly rapidly with wavelength longward of 1.3 cm. Visible and IR spectroscopy show that NH_3, whose strong inversion spectrum peaks at ~1.3 cm, is an important tropospheric species at Jupiter and Saturn. Its signature on the Jovian radio spectrum is obvious, causing the prominent "hole" at 1.3 cm. At Saturn it is more subdued but is the source of that spectrum's change in slope at 1.3 cm. Radiative transfer models of Jupiter and Saturn with near-solar deep NH_3 abundances agree well with the data (*e.g.*, de Pater, 1990).

Uranus' T_B spectrum does not fit this pattern. T_B is ~175 K at 1.3 cm, ~80 K warmer than at 0.1 cm and much warmer than the other three planets. The data indicate a nearly linear increase in T_B with $\log(\lambda)$ over the entire range shown in figure 3, requiring no break in slope near 1.3 cm. Notably, there is a distinct difference between Uranus' and Neptune's spectra: at ~20 cm, and also at 0.1-0.4 cm, Uranus' T_B are quite similar to Neptune's, but in the intervening 1-10 cm range Uranus averages 30-55 K warmer than Neptune. Gulkis *et al.* (1978) first showed that Uranus radiative transfer models with near-solar NH_3 deep abundances predict T_B at cm wavelengths that are much too cold. Their model using an NH_3 abundance about 1% of solar fit the data better, though far

Earth, Moon, and Planets **67**: 89–94, 1995.
© 1995 *Kluwer Academic Publishers.*

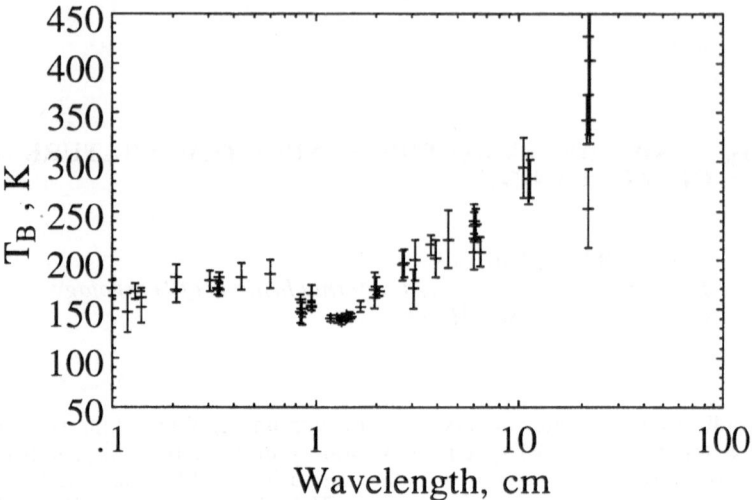

Figure 1. Jupiter's radio brightness temperature spectrum

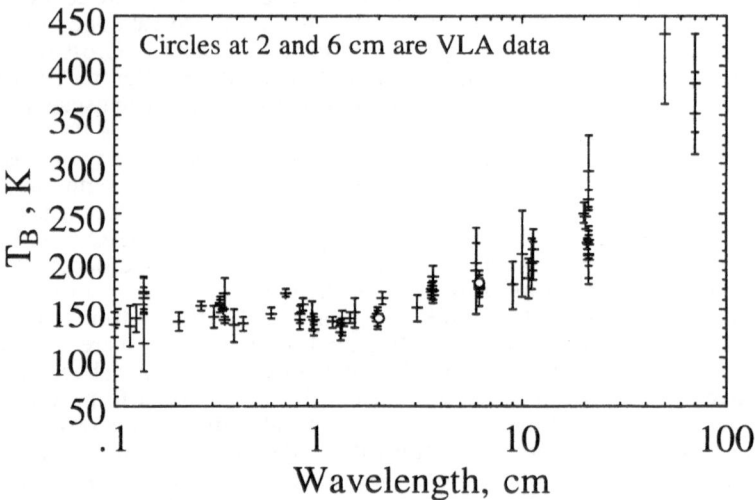

Figure 2. Saturn's radio brightness temperature spectrum

from perfectly. They offered one possible cause for the apparent NH_3 depletion: a superabundance of H_2S could react out most of the NH_3, forming NH_4SH as first discussed by Wildt (1937). At the time most researchers assumed that essentially all nitrogen in giant planet tropospheres is in the highly reduced form. In that case the solar

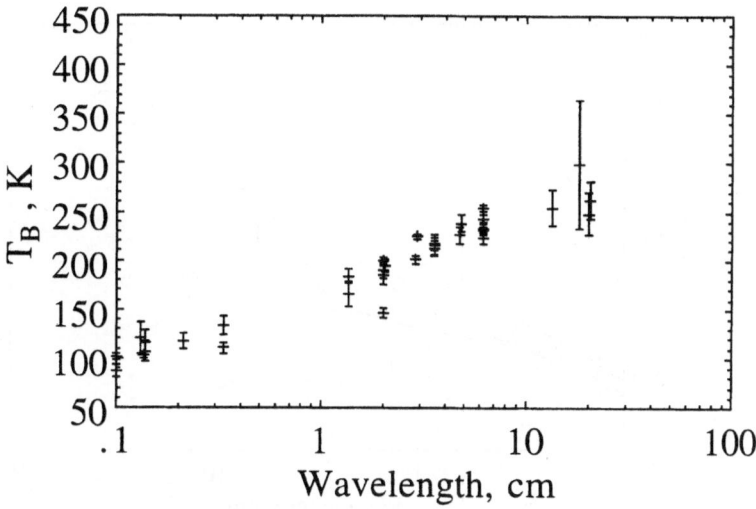

Figure 3. Uranus' radio brightness temperature spectrum

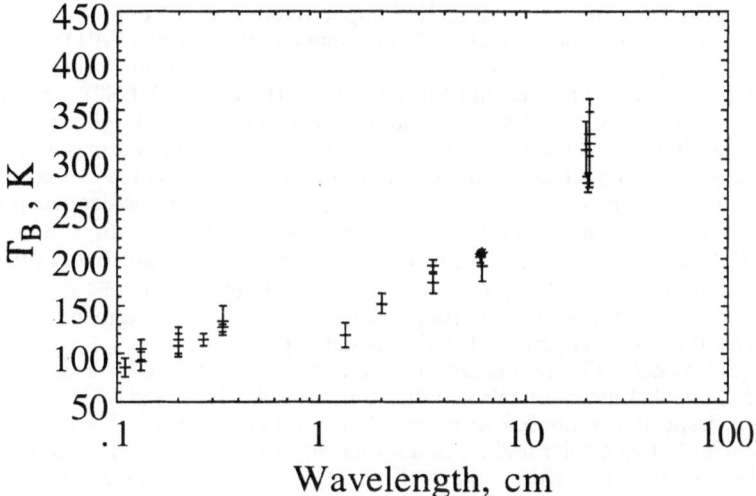

Figure 4. Neptune's radio brightness temperature spectrum

abundance of sulfur made H_2S seem the most likely candidate to deplete NH_3, despite the lack of direct observational evidence of H_2S at Uranus or any of the other giant planets. This nondetection is expected even if sulfur abundances are approximately solar, since models predict the NH_4SH-forming reaction would restrict H_2S to depths inaccessible to visible or IR spectroscopy (Atreya and Romani, 1985).

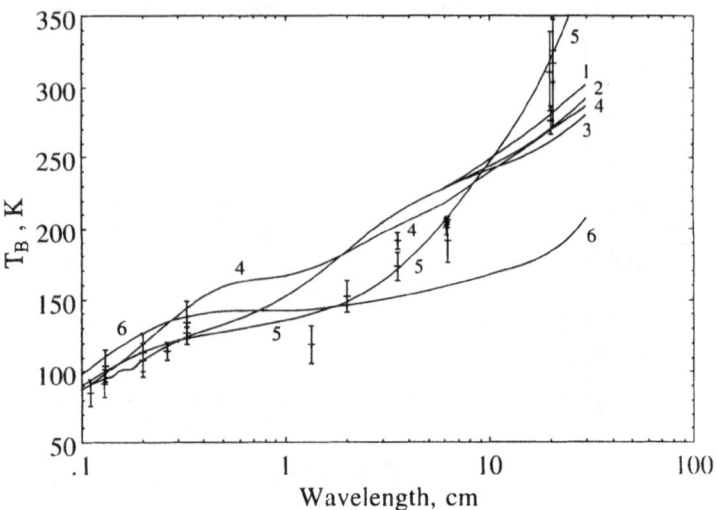

Figure 5. Neptune's radio brightness temperature spectrum, with predictions of various models superposed. See text for model descriptions.

Some researchers suggest a large H_2S superabundance at Neptune also. Problems fitting radiative transfer models to cm data prompted de Pater *et al.* (1991) to invoke NH_3-depleting H_2S at Neptune, and to suggest that H_2S might contribute significantly to the total opacity. Recently DeBoer and Steffes (1994) (hereafter "DBS") made laboratory measurements of cm H_2S opacities and found them a factor of two larger than Van Vleck-Weisskopf predictions. Based on this they suggest H_2S may be the major source of cm opacity in Neptune's upper troposphere, and reinterpret Lindal's (1992) Voyager 2 radio occultation data. Lindal assumed all opacity at the 6.3 bar level, the deepest probed, was due to NH_3 and derived a volume mixing ratio of 5×10^{-7}. DBS assume all opacity there is due to H_2S and derive a mixing ratio of 1.7×10^{-4}. Based on that result, they use radiative transfer models with H_2S above the NH_4SH cloud to generate predicted spectra they compare to the Neptune T_B data (both single-dish and VLA). Agreement between the predicted spectra and the superior VLA data is rather poor.

Figure 5 duplicates Figure 4 except it includes results from various radiative transfer models. Models 1-4 are after DBS, with 30 times solar H_2O and CH_4; NH_3 and H_2S abundances, respectively, are 0.5 solar and 15 times solar for model 1, solar and 18 times solar for model 2, twice solar and 25 times solar for model 3, and solar and 6 times solar for model 4. Models 1-3, whose spectral results are identical shortward of 6 cm, yield 1.7×10^{-4} H_2S above the NH_4SH cloud, while model 4 yields Lindal's 5×10^{-7} NH_3. Model 5 is after de Pater and Richmond (1989), using an ~2% solar NH_3 mixing ratio (3×10^{-6}) throughout the atmosphere, limited by saturation, and no H_2S. Model 6, by the author, uses approximately solar NH_3 (2×10^{-4}) and no H_2S to demonstrate that Neptune models with uniformly near-solar tropospheric NH_3 mixing ratios are not consistent with the observed spectrum, so some form of NH_3 depletion (not necessarily H_2S) is needed.

Only the models with NH_3 above the NH_4SH cloud reproduce Neptune's T_B dip at cm wavelengths. Model 5, with more NH_3 than model 4, provides the best fit; slightly more NH_3 would provide a better fit, further decreasing T_B longward of 1 cm. This does not conflict with Lindal's result, since he states NH_3 is probably still saturated at the deepest datum. Due to the H_2S spectrum's simple f^2 dependency longward of 0.4 cm, models dominated by H_2S above the NH_4SH cloud (DBS models 1-3) deviate <10 K from a straight line on the plot, quite unlike the data. Reproducing the T_B dip with such an absorber requires a relatively thin tropospheric layer near the 120-130 K level (~5 bar pressure level) with a much larger absorber mixing ratio than adjacent layers, a situation more appropriate to the upper stratosphere than the troposphere. Without a mechanism to maintain such a tropospheric layer, it is unlikely that the observed cm opacity in Neptune's upper troposphere is primarily due to H_2S. Neptune's radio spectrum appears to require NH_3, or another species with an opacity peak near 1-2 cm, in the upper troposphere.

Applying the DBS models 1-3 to Uranus leads to a different conclusion for that planet. Upper tropospheric T-P (temperature-pressure) relations for Uranus and Neptune are very similar: at equal pressures, their temperatures differ by ~5K at most from well above their

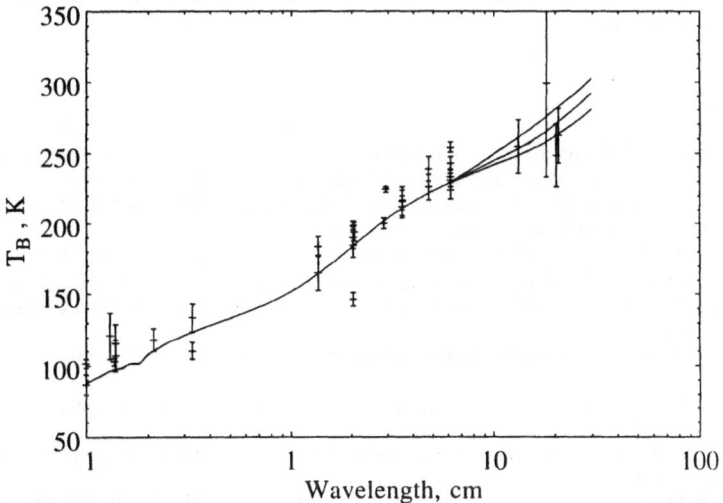

Figure 6. Uranus' radio brightness temperature spectrum, with predictions of H_2S-dominated models of Neptune by DBS (solid lines). See previous page for model descriptions.

tropospheres to the deepest level probed by radio occultation (Lindal *et al.*, 1987; Lindal, 1992). Also, recent work by Killen and Flasar (1995) show that the cm spectra of the giant planets are largely insensitive to T-P profiles, but instead depend most strongly on the relative humidities of condensible absorbers. Given a fixed set of constituent abundance profiles, a model using Neptune's T-P profile will yield a T_B spectrum quite similar to one produced using Uranus' T-P profile. The dissimilarity of the two planets' observed radio spectra makes it highly unlikely they have similar constituent profiles. Figure 6 shows the result of using DBS' H_2S-dominated models of Neptune as first approximations to such models for Uranus. The models fit Uranus' observed spectrum much better than Neptune's, suggesting that tropospheric constituents whose cm opacities have f^2 dependencies, such as H_2S, could produce Uranus' observed radio T_B spectrum.

This does not rule out the presence of NH$_3$, but it points to other possibilities. Thus, while Neptune seems to need a small (relative to solar) but nontrivial amount of NH$_3$ in its upper troposphere, Uranus does not.

Recent work by Atreya *et al.* (1995) suggests that a significant fraction of Uranus' and Neptune's nitrogen could be in the form of N$_2$ instead of NH$_3$. Thus a range of nitrogen and sulfur abundances, both not greatly different from solar, could yield the observed depletion of NH$_3$. Further radiative transfer modeling can better define the limits of such abundances, which currently are poorly constrained. Insufficient data are available at this time to state with certainty the precise mechanisms for NH$_3$ depletions at Uranus and Neptune. Whatever the mechanisms, comparing the two planets' radio T$_B$ spectra shows that the resulting absorber mixing ratio profiles in their upper tropospheres are quite different.

Acknowledgement

The research described in this paper was carried out at the Jet Propulsion Laboratory, California Institute of Technology, under a contract with the National Aeronautics and Space Administration.

REFERENCES

Atreya, S.K., and P.N. Romani (1985) Photochemistry and clouds of Jupiter, Saturn, and Uranus, in *Recent Advances in Planetary Meteorology*, G.E. Hunt ed., Cambridge Univ. Press, Cambridge, 17-68.

Atreya, S.K., S.G. Edgington, D. Gautier, and T.C. Owen (1995) Origin of the Major Planet Atmospheres: Clues from Trace Species, this publication.

DeBoer, D.R., and P.G. Steffes (1994) Laboratory Measurements of the Microwave Properties of H$_2$S under Simulated Jovian Conditions with an application to Neptune, *Icarus* **109**, 352-366.

de Pater, I., and S.T. Massie (1985) Models of the Millimeter-Centimeter Spectra of the Giant Planets, *Icarus* **62**, 143-171.

de Pater, I., and M. Richmond (1989) Neptune's Microwave Spectrum from 1 mm to 20 cm, *Icarus* **80**, 1-13.

de Pater, I. (1990) Models of the Millimeter-Centimeter Spectra of the Giant Planets, *Ann.Rev.Astr. Astphy* **28**, 347-399.

de Pater, I., P.N. Romani, and S.K. Atreya (1991) Possible Microwave Absorption by H2S Gas in Uranus' and Neptune's Atmospheres, *Icarus* **91**, 220-233.

Griffin, M.J., and G.S. Orton (1993) The Near-Millimeter Brightness Temperature Spectra of Uranus and Neptune, *Icarus* **105**, 537-547.

Grossman, A.W., D.O. Muhleman, and G.L. Berge (1989) High-Resolution Microwave Images of Saturn, *Science* **245**, 1211-1215.

Gulkis, S., M.A. Janssen, and E.T. Olsen (1978) Evidence for the Depletion of Ammonia in the Uranus Atmosphere, *Icarus* **34**, 10-19.

Hofstadter, M.D. (1993) Microwave Imaging of Neptune's Troposphere, *Bull.Am.Astr.Soc.* **25** No. 3, 1077.

Killen, R.M., and F.M. Flasar (1995) Microwave Sounding of the Giant Planets, *Icarus*, in press.

Lindal, G.F., J.R. Lyons, D.N. Sweetnam, V.R. Eshleman, D.P. Hinson, and G.L. Tyler (1987) The Atmosphere of Uranus: Results of Radio Occultation Measurements With Voyager 2, *JGR* **92** No. A13, 14987-15001.

Lindal, G.F. (1992) The Atmosphere of Neptune: An Analysis of Radio Occultation Data Acquired With Voyager 2, *Astron.J* **103** No. 3, 967-982.

Muhleman, D.O., and G.L. Berge (1991) Observations of Mars, Uranus, Neptune, Io, Europa, Ganymede, and Callisto at a Wavelength of 2.66 mm, *Icarus* **92**, 263-272.

TITAN'S ATMOSPHERE AND SURFACE: PARALLELS AND DIFFERENCES WITH THE PRIMITIVE EARTH

A. COUSTENIS
Dept. of Space Research
Paris-Meudon Observatory, 92195 Meudon, France

Abstract. In spite of a marked resemblance with our planet, Titan should not be hastily considered as another Earth but rather as a useful tool in the study of chemical and physical processes in the primitive Earth.

1   Introduction

There exist only 3 or 4 planetary objects that have N_2 atmospheres in our Solar System (Earth, Titan, Triton and perhaps Pluto). The present state and chemical evolution of the atmospheres of Titan, Triton and Pluto is explained in Lunine, Atreya and Pollack (1989) and references therein. For purposes of comparative planetology, I have compiled from the various sources referenced in this paper Tables 1 to 4. This article, however, is mainly focused on Titan vs Earth comparisons. Because Titan has a substantial atmosphere with molecular nitrogen as the major component, the satellite is evidently the best laboratory available to scientists at this time for studying the chemical evolution of an atmosphere similar to that of the primitive Earth. For an extended review of the satellite's characteristics, see Hunten *et al.* (1984). Although the satellite is half our planet's size (Table 1), its atmosphere is denser and larger than the Earth's reaching out in the space up to about 1500 km from the surface (Table 2). Other properties of Titan were more precisely determined following the Voyager 1 encounter: the total mass of the satellite is about 1.4×10^{26} g, or 1/50 that of the Earth's; the surface gravity is about 7 times less than on Earth; its obliquity is assumed to be 26.7, while the Bond albedo of the satellite is estimated to be around 0.25; at a distance of 1.3 $\times 10^6$ AU from the Sun, Titan receives only 1.1% of the solar flux that reaches our Earth.

2   Parallels between Titan and the primitive Earth

Many similarities, discussed hereafter, between Titan and our planet, can be considered as arguments in favor of regarding Ttian as a good early Earth analogy.

Earth, Moon, and Planets **67**: 95–100, 1995.
© 1995 *Kluwer Academic Publishers.*

TABLE 1. Planetary and Satellite Orbital Data

	Titan	Earth	Triton	Pluto
Semi-major axis of orbit (AU)	9.54	1	30.06	39.44 (29.7 now)
Radius (km)	2575	6378	1352	1208?
Rotation period (day)	16	1	5.9	6.4
Sidereal period (yr)	29.46	1	165	248
Surface gravity ($cm\ s^{-2}$)	135	979	78.1	59.7

TABLE 2. Atmospheric Structure/Temperature Data

	Titan	Earth	Triton	Pluto
Surface temperature (K)	94	288	38	35-55
Albedo	0.2-0.3	0.3	0.7	0.2-1.
Surface Pressure (bar)	1.45	1	14×10^{-6}	3×10^{-6} ?
Temperature 1 mbar (K)	150-170	190	41	104 ± 21
Tmax (K)	185 ± 20	600-1500	102 ± 3	?
$\lambda = r/H$ $= mgr/kT$ at 1 mbar	68	1140	84	21
Exobase height (km)	1500	500	930	?
$r_{exobase}/r_{planet}$	1.6	1.08	1.7	>>3 ?

Cautiousness is advised, however, in hastily subscribing to this analogy, because of the important differences which also exist, as explained in Section 3.

2.1 THE ATMOSPHERE: THERMAL AND CHEMICAL COMPOSITION

Nine times farther from the Sun than our planet, Saturn's largest moon exhibits

TABLE 3. Chemical Composition of Atmospheres

		Titan	Earth	Triton	Pluto
Major constituent (%)	N_2	90-97	78	99	99 ?
Minor species (%)	O_2		21		
	H_2O	? ($<10^{-9}$)	0.1-3	?	?
	Ar	? (<10)	0.93	?	?
	CO_2	10^{-8}	0.034	?	?
	CO	10^{-6}-10^{-4}	10^{-5}	2×10^{-4}	?
	CH_4	2-3 (strat.)	1.5×10^{-4}	10^{-4}	?
	H_2	0.2		10^{-4}	?

a thermal structure much like the Earth's but much colder (the solar radiation reaching Titan's haze top is about 100 times less than on Earth). The effective temperature at which Titan radiates to space is 83°K and corresponds to about 73% absorption of the incident solar energy in the upper atmosphere. The temperature decreases in the stratosphere as one descends from 600 to 40 km of altitude (Lellouch et al.,1989) and reaches the tropopause, where the minimum value of 71°K is attained (Lindal et al.,1983). There, a temperature inversion occurs, as a moderate greenhouse effect heats up the surface to 94°K for a pressure of 1.5 bar (Table 2). The meridional temperature distribution at the time of the Voyager encounter (spring equinox) was not symmetric about the equator and latitudinal variations show the north polar region to be 15-20°K colder than the equator or the south polar regions (Flasar and Conrath 1990). This implies that Titan's atmosphere may be controlled by dynamical inertia, to account for the atmospheric lag behind the solar input, or/and that the temperature asymmetry could be due to compositional variations (Coustenis and Bézard 1994).

Both our planet and Titan support active organic chemistry in their atmospheres (Earth is also biologically controlled). After N_2, CH_4 is the most abundant species on Titan (0.5-3.4% in the stratosphere), followed by H_2 and others (Table 3). Carbon-containing species and nitriles exist on Titan in detectable quantities. Voyager 1 and ground-based measurements have detected hydrocarbons (ethane, acetylene, propane, etc...) in complex forms, nitriles (HCN, HC_3N, C_2N_2, CH_3CN...) - some of which are prebiotic molecules -, and two oxygen compounds (CO, CO_2). However, the latter are scarce, and oxygen exists on Titan only in trace amounts (Table 3). Water has not yet been detected. The minor species in Titan's atmosphere were found to show latitudinal variations, much like on Earth, which could be due to seasonal effects (Coustenis and Bézard 1994).

2.2 ATMOSPHERIC CONDENSATES AND SURFACE

The orange-yellow haze that globally covers the satellite suggests the presence of aerosols in its atmosphere, as on Earth. The nature and size of the aerosols is not yet well defined, but we know that the molecular fragments and other components produced by photolysis and energetic electrons in the upper part of the atmosphere form polymers, as suggested by laboratory studies (Sagan et al.,1984). All the hydrocarbons and nitriles produced by N_2-CH_4 reactions (Yung et al.,1984) condense in the coldest region of the atmosphere and form solid suspended particles which probably sink lower into the deeper atmosphere, collide and form aggregates dropping even faster to finally land on the surface of the satellite, where they may rest or be absorbed by a porous surface material and end up stored in a reservoir underneath the "ground". Methane clouds, instead of water vapor ones, may exist near the tropopause, although it has been suggested recently that supersaturation of methane could occur, leading to a partially aerosol-free troposphere instead (Samuelson 1994, Courtin et al.,1994).

Recent observations of Titan put a question mark on previous theoretical models which suggested the presence of large liquid hydrocarbon reservoirs on the surface (Lunine et al.,1983). The mystery remains unsolved, but efforts to reconcile observational data and theory set forth various possibilities which include a frothy aerosol-contaminated ocean, a non-global smaller ocean, or parts of dry land and lakes (Lunine 1993). A mostly solid surface is likely based on radar measurements (Muhleman et al.,1990) and on near-infrared observations compiled in Table 4, (Lemmon et al.,1993; Griffith 1993), perhaps covered with dirty water ice, or with a porous icy or rocky regolith in which the hydrocarbon reservoir is stored (Stevenson 1992; Coustenis et al.,1994).

3 Differences with the primitive Earth

Organic gases, aerosol layers, condensation and perhaps methane clouds exist in Titan's atmosphere, and a liquid ocean (or lakes) as well as ices and minerals have

TABLE 4. Surface Composition

	Titan	Earth	Triton	Pluto
Ocean	Hydrocarbons (C_2H_6-CH_4-N_2)	H_2O	Frozen	Frozen
Ices	C_2H_2,CO_2, H_2O, ?	H_2O	N_2,CO, CH_4,CO_2	N_2, CO, CH_4
Crust	dirty ice, minerals?	minerals		

been suggested to cover its surface. However, conditions on present day Titan do not replicate with precision those on primordial Earth and some of the reasons are the following. *(a)*: The major composition on Titan (N_2-CH_4) is different from the Earth's (N_2-CO-CO_2). *(b)*: The low temperature on Titan, due to less visible radiation (100 times) than that reaching the Earth, certainly slows the chemical reactions considerably and may have up to now prevented the formation of complex molecules required in the life-chain. *(c)*: Water, although suspected in solid form, has not yet been firmly detected. *(d)*: The state of oxidation between the inner and the outer solar system atmospheres is very different. C is present in its fully oxidized state, CO_2, in all the terrestrial planets and in its fully reduced state in most of the outer Solar System objects. This difference could reflect the ability of water to buffer the oxidation state of the carbon species in the atmospheres of the terrestrial planets, while the oxidation state of the C in the outer Solar System atmospheres may be primordial (Lunine *et al.*,1989). The present-day Earth is an extreme case in this regard due to the high abundance of O_2. This apparently reflects the Earth's unique position in the solar system as an abode of life. *(e)* The infall of carbonaceous material (meteorites, comets, etc...) on a planet, is smaller today than in the past. *(f)* the Earth being larger than Titan, its atmosphere has evolved significantly over its lifetime due to volcanic and tectonic changes with time. Titan, by way of contrast, may have undergone early dramatic changes followed by tectonic quiescence. Subsequent evolution would be externally driven by photochemical processes perhaps by tectonical activity (Lunine and Stevenson 1987).

4   Conclusions and perspectives

All this may signify that the satellite will never be more than a good model for the early Earth, but Titan is undeniably a valuable tool to the study of some chemical and physical processes. The *Cassini/Huygens* mission to Saturn and Titan (to be launched in 1997) will arrive at Titan in 2004 to make orbital and in situ measurements for four years, bringing new insight to the Earth-Titan resemblances/differences. In the expectation of the wealth of data from this mission, astronomers will still use all available means from the ground and satellites (such as *ISO, Infrared Space Observatory*) to study this fascinating and promising object.

5   References

Atreya, S.K. (1986) *Atmospheres and Ionospheres of the Outer Planets and Their Satellites*, Springer-Verlag (eds.), New York, 145-197.

Coustenis, A., and Bézard, B. (1995) Titan's atmosphere from Voyager infrared observations. IV. Latitudinal variations of temperature and composition. *Icarus* , in press.

Coustenis, A., Lellouch, E., Maillard J.P., and McKay, C.P. (1995) Titan's surface: composition and variability from its near infrared albedo. Submitted to *Icarus*.

Courtin, R., Gautier, D., and McKay, C.P. (1994) Titan's thermal emission spectrum: re-analysis of the Voyager infrared measurements. *Bull. Am. Astron. Soc.* **26**, 1182.

Flasar, F.M., and Conrath, B.J. (1990) Titan's stratospheric temperatures: a case for dynamical inertia? *Icarus* **85**, 346-354.

Griffith, C.A. (1993) Evidence for surface heterogeneity on Titan. *Nature* **364**, 511-514.

Hunten, D.M., Tomasko, M.G., Flasar, F.M., Samuelson, R.E., Strobel, D.F., and Stevenson, D.J. (1984) Titan. In *Saturn*, T. Gehrels and M.S. Matthews (eds.), University of Arizona Press, Tucson, 671-759.

Lellouch, E., Coustenis, A., Gautier, D., Raulin, F., Duboulos, N., and Frère, C. (1989) Titan's atmosphere and hypothesized ocean: A re-analysis of the Voyager 1 radio-occultation and IRIS 7.7 μm data. *Icarus* **79**, 328-349.

Lemmon, M.T., Karkoschka, E., and Tomasko, M. (1993) Titan's rotation: surface feature observed. *Icarus* **103**, 329-332.

Lindal, G.F., Wood, G.E., Hotz, H.B., Sweetnam, D.N., Eshelman, V.R., and Tyler, G.L. (1983) The atmosphere of Titan: an analysis of the Voyager 1 radio-occultation measurements. *Icarus* **53**, 348-363.

Lunine, J.I. (1993) Does Titan have an ocean? A review of current understanding of Titan's surface. *Rev. of Geophys.* **31**, 133-149.

Lunine, J.I., Stevenson, D.J., and Yung, Y.L. (1983) Ethane ocean on Titan. *Science* **222**, 1229-1230.

Lunine, J.I., and Stevenson, D.J. (1985) Evolution of Titan's coupled ocean-atmosphere system and interaction of ocean with bedrock. In *Ices in the Solar System*, J. Klinger, D. Benest, A. Dollfus, and R. Smoluchowski (eds.), D. Reidel (Publ.), Dordrecht, 741-757.

Lunine, J.I., and Stevenson, D.J. (1987) Clathrate and ammonia hydrates at high pressure: application to the origin of methane on Titan. *Icarus* **70**, 61-77.

Lunine, J.I., Atreya, S.K., and Pollack, J.B. (1989) Present state and chemical evolution of the atmospheres of Titan, Triton, and Pluto. In *Origin and Evolution of Planetary and Satellite Atmospheres*, S.K. Atreya, J.B. Pollack, and M.S. Matthews (eds.), The Univ. of Arizona Press, Tucson, 605.

Muhleman, D.O., Grossman, A.W., Butler, B.J., and Slade, M.A. (1990) Radar reflectivity of Titan. *Science* **248**, 975-980.

Sagan, C., Khare, B.N., and Lewis, J.S. (1984) Organic matter in the Saturn system. In *Saturn*, T. Gehrels and M.S. Matthews (eds.), Univ. of Arizona Press, Tucson, 788-807.

Samuelson, R.E. (1994) Methane supersaturation in Titan's atmosphere. *Bull. Am. Astron. Soc.* **26**, 1183.

Stevenson, D.J. (1992) Interior of Titan. *ESA SP-338*, ESTEC, Noordwijk, The Netherlands, 29-33.

Y.L., Allen, M., and Pinto, J.P. (1984) *Astrophys. J. Supp.* **55**, 465-506.

CRYOVOLCANISM ON THE ICY SATELLITES

J.S. KARGEL
U.S. Geological Survey
2255 N. Gemini Dr., Flagstaff, AZ 86001

Abstract

Evidence of past cryovolcanism is widespread and extremely varied on the icy satellites. Some cryovolcanic landscapes, notably on Triton, are similar to many silicate volcanic terrains, including what appear to be volcanic rifts, calderas and solidified lava lakes, flow fields, breached cinder cones or stratovolcanoes, viscous lava domes, and sinuous rilles. Most other satellites have terrains that are different in the important respect that no obvious volcanoes are present. The preserved record of cryovolcanism generally is believed to have formed by eruptions of aqueous solutions and slurries. Even Triton's volcanic crust, which is covered by nitrogen-rich frost, is probably dominated by water ice. Nonpolar and weakly polar molecular liquids (mainly N_2, CH_4, CO, CO_2, and Ar), may originate by decomposition of gas-clathrate hydrates and may have been erupted on some icy satellites, but without water these substances do not form rigid solids that are stable against sublimation or melting over geologic time. Triton's plumes, active at the time of Voyager 2's flyby, may consist of multicomponent nonpolar gas mixtures. The plumes may be volcanogenic fumaroles or geyserlike emissions powered by deep internal heating, and, thus, the plumes may be indicating an interior that is still cryomagmatically active; or Triton's plumes may be powered by solar heating of translucent ices very near the surface. The Uranian and Neptunian satellites Miranda, Ariel, and Triton have flow deposits that are hundreds to thousands of meters thick (implying highly viscous lavas); by contrast, the Jovian and Saturnian satellites generally have plains-forming deposits composed of relatively thin flows whose thicknesses have not been resolved in Voyager images (thus implying relatively low-viscosity lavas). One possible explanation for this inferred rheological distinction involves a difference in volatile composition of the Uranian and Neptunian satellites on one hand and of the Jovian and Saturnian satellites on the other hand. Perhaps the Jovian and Saturnian satellites tend to have relatively "clean" compositions with water ice as the main volatile (ammonia and water-soluble salts may also be present). The Uranian and Neptunian satellites may possess large amounts of a chemically unequilibrated comet-like volatile assemblage, including methanol, formaldehyde, and a host of other highly water- and ammonia-water-soluble constituents and gas clathrate hydrates. These two volatile mixtures would produce melts that differ enormously in viscosity The geomorphologic similarity in the products of volcanism on Earth and Triton may arise partly from a rheological similarity of the ammonia-water-methanol series of liquids and the silicate series ranging from basalt to dacite. An abundance of gas clathrate hydrates hypothesized to be contained by the satellites of Uranus and Neptune could contribute to evidence of explosive volcanism on those objects.

Earth, Moon, and Planets **67**: 101–113, 1995.
© 1995 *Kluwer Academic Publishers.*

I. Introduction

In six planetary flybys during the period from 1979-1989 the two Voyager spacecraft unveiled the surfaces of numerous satellites of the four largest planets in the solar system. Voyager images showed for the first time details of the ice-covered surfaces and the dynamic geologic histories of many of these objects (Smith *et al.*, 1979, 1982, 1986, 1989). The surfaces of the icy satellites present almost as great a range of geologic phenomena as presented by the rocky planets (Rothery 1992), though we know much less about the former than the latter. The basic properties of icy satellites are quite varied (Burns 1986). Their unique surface expressions apparently reflect individualized geologic evolutionary histories that, in turn, relate to some very large differences in factors such as thermal evolution, composition, interior pressure, surface temperature, surface gravity, and impact history.

None of the very small, irregularly shaped icy satellites of the outer solar system (such as Proteus and Hyperion and other satellites with mean radii less than ~200 km) have any indication of resurfacing except by the impacts that formed them as fragments of once-larger objects (Thomas 1989, Farinella *et al.* 1990, Croft 1992). Some major ellipsoidally shaped icy satellites, notably Mimas, Rhea, and Callisto, also lack compelling evidence of volcanic resurfacing; instead, they have intensely cratered surfaces everywhere that the Voyager spacecraft looked (Smith *et al.*, 1979, 1982). In a dramatic confirmation of pre-Voyager predictions that large icy satellites should be differentiated (Lewis 1971, Consolmagno and Lewis 1978), the Voyager project found that some of the most interesting geologic processes on many icy satellites involved cryovolcanism (i.e., the eruption of icy-cold aqueous or nonpolar molecular solutions or partly crystallized slurries, derived by partial melting of ice-bearing materials) (Smith *et al.*, 1979, 1982, 1986, 1989). What was not predicted was the discovery of active cryovolcanism or gas venting on Triton and evidence of past cryovolcanism on objects as small as Enceladus and Miranda (with radii of 249 and 242 km, respectively). Several active, surface-vented plumes were indeed observed on Triton during the last of the Voyager satellite encounters (Smith *et al.* 1989, Kirk *et al.* 1995). Other satellites, notably Enceladus and Europa, have strong geologic evidence of geologically recent and possibly (but not necessarily) active volcanic resurfacing (Smith *et al.* 1979, 1982; Lucchitta and Soderblom 1982; Squyres *et al.* 1983). Many other icy satellites, notably Ganymede, Tethys, Dione, Miranda, and Ariel, have abundant evidence of ancient cryovolcanic resurfacing, but virtually no evidence for ongoing or geologically very recent activity (Smith *et al.* 1979, 1982, 1986; Shoemaker *et al.* 1982; Croft and Soderblom 1991).

Unfortunately, very little is known of the exact fluids that were responsible for the genesis of the varied cryovolcanic landscapes. Indirect evidence suggests that the lavas were fundamentally aqueous liquids or partly crystallized aqueous slurries in most instances, though they may have had considerable chemical variability (Stevenson 1982; Squyres *et al.* 1983; Smith *et al.*, 1989, Kargel and Strom 1990, Croft and Soderblom 1991; Kargel 1991, 1992; Kargel *et al.* 1991; Schenk 1991), and some of these "lavas" possibly may even have included eruptions of water mainly in the solid state (Jankowski and Squyres 1988). The exact compositions of the liquids generated by partial melting in the interiors of icy satellites depends on the composition of the ices and of the water-soluble material in the rocky fraction. Until we obtain actual samples of icy satellites, our knowledge of their compositions is underconstrained by a few critical observations and theoretical calculations.

II. Cryovolcanic landforms and landscapes

The resolution of the best Voyager images of icy satellites was commonly sufficient to resolve features at a scale of 1-2 km, good enough to identify volcanic terrains but generally not good enough to resolve details of individual flow units unless they are of very great thickness (as they are on Miranda, Ariel, and Triton). Under favorable conditions of imaging and illumination, for instance near the terminator, tectonic scarps and crater rims on Ganymede and Enceladus that are ~100 m high are readily visible in Voyager images, yet good examples of flows of comparable (or greater) thickness were not found on either object (or on any other satellite in the Jovian and Saturnian systems). The extrusional origin of cryovolcanic plains on Ganymede and Enceladus was recognized by the similarity of their appearance and stratigraphic relations to plains of known volcanic origin-- the lunar maria, for instance, consist of smooth deposits of basaltic lava that (1) occupy low-lying regions, (2) embay or partly bury large impact craters and other types of adjoining rugged highlands, and (3) apparently completely bury large numbers of small impact craters (inferred from crater size-frequency statistics; Strom 1986). These relations are visible in images of the lunar maria at resolutions comparable to those produced by Voyager, but individual lava flows are not recognizable. Some of the plains on Ganymede can be traced to specific fracture zones, which probably served as eruption conduits from which vast sheets or lakes of water or aqueous solutions apparently were erupted. Hence, there is little doubt that the smooth resurfaced plains on many icy satellites are actually of extrusional origin.

At the simplest level of interpretation, effusive cryovolcanic deposits come in two types: (1) plains-forming flows of unresolved relief occur on several Jovian and Saturnian satellites, including Europa (Fig. 1), Ganymede (Fig. 2), Enceladus (Fig. 3), Tethys, and Dione, as well as Neptune's satellite, Triton; and (2) thick, lobate flow units (hundreds to a couple thousand meters thick), large volcanic constructs, and volcanic channels occur on the Uranian and Neptunian satellites, including Miranda (Fig. 4), Ariel (Fig. 5), and Triton (Figs. 6-8).

Some of the most interesting cryovolcanic features are those where individual flow units and other discrete volcanic landforms can be resolved, especially the volcanic features of Triton, which has cryovolcanic landscapes that are most astonishing for their similarity to volcanic terrains on Earth and the other rocky bodies (Kargel and Strom 1990, Croft *et al.* 1995). Triton has features that are morphologically similar to volcanic calderas and associated flow fields on Venus, Earth, Mars, and Io; rhyolite and dacite flows and domes on Earth; cinder cones and other cratered volcanic cones on Venus and Earth; rift-controlled crater chains on Venus, Earth, and Mars; and crater-headed sinuous rilles or collapsed lava tubes on Venus, Earth, Moon, and Mars. This similarity seems to require cryolavas on Triton that behaved volcanologically much as silicate lavas behave on the terrestrial planets.

Evidence for explosive cryovolcanism is widespread on some icy satellites, especially on Triton (also possibly on Enceladus and Miranda), but absent on other satellites (notably Europa and Ganymede). Triton has cratered cones that resemble breached cinder cones and composite stratovolcanoes (such as Mount Saint Helens) on Earth. Each of Triton's large smooth-floored walled planitia (caldera or lava lake analogs) has a central hummocky and pitted region that may have been the site of late-stage cryoclastic eruptive activity. Elsewhere on Triton, a mantling deposit tapers to a feather edge, suggesting that this terrain is overlain by a cryoclastic blanket analogous to terrestrial ignimbrite sheets.

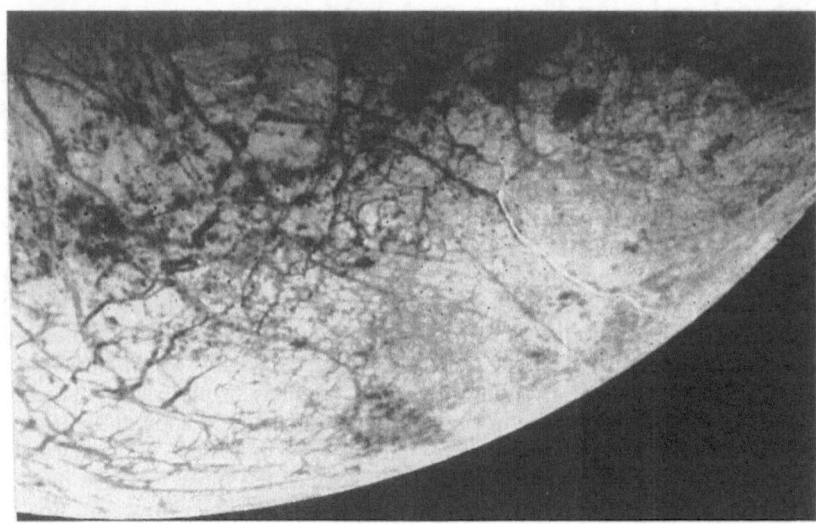

Figure 1. Voyager image 20649.25, showing a smooth but highly fractured surface of Europa. This scene has been likened to sea ice, and it suggests complete flooding and tectonic disruption of the surface. Scene ~1600 km tleft to right .

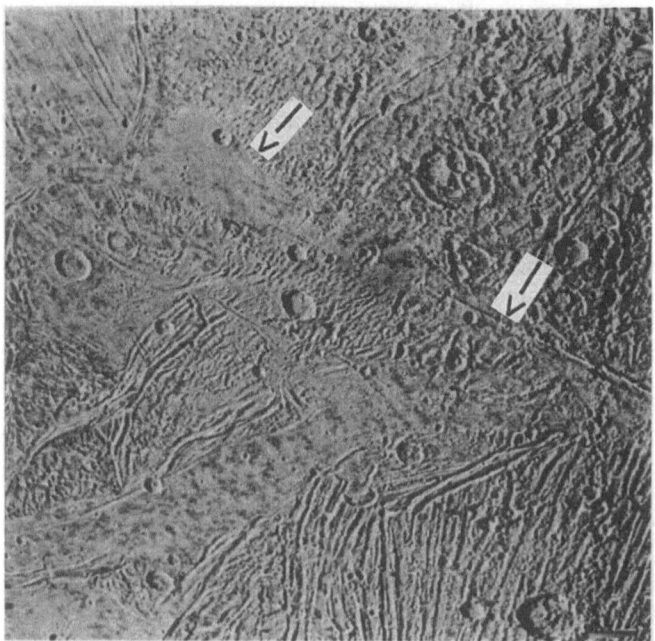

Figure 2. Voyager image 20640.25, showing a typical region of Ganymede's young, lightly cratered grooved terrain and remnants of an older cratered terrain. Groove sets are locally completely flooded by smooth plains composed of cryovolcanic deposits of higher albedo than the cratered terrain (albedo differences are suppressed in this processed version of the image). The smooth terrain in the upper left quadrant (arrow) appears to have been extruded from a prominent fracture zone visible at right center (arrow). Scene ~850 km across.

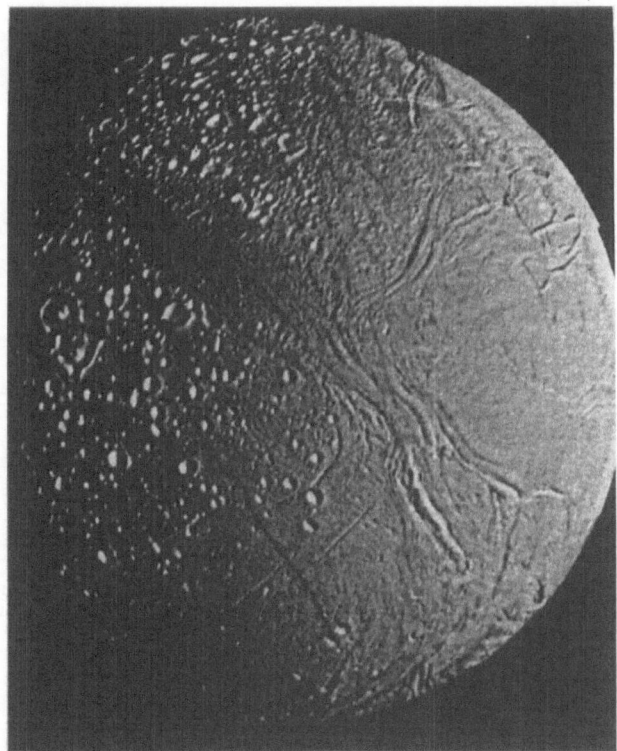

Figure 3. Global view of Enceladus (diameter 500 km). Sinuous mountain ridges, over 1 km high in places, and fracture belts cut areas of lightly to heavily cratered plains. Smooth plains appear to be cryovolcanic.

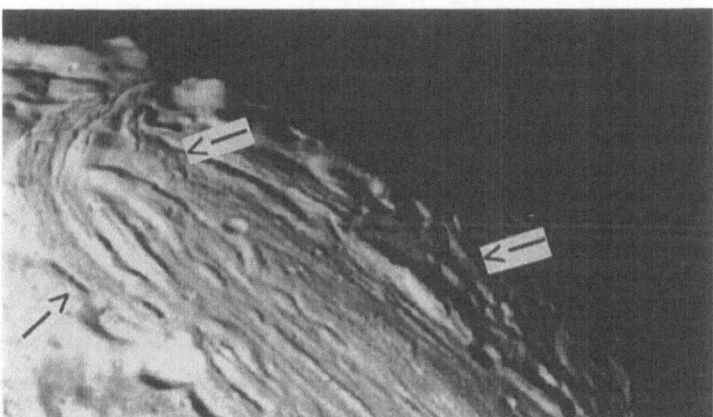

Figure 4. Image showing cryovolcanic plains in part of Elsinore Corona, Miranda, apparently including highly viscous effusive deposits (arrows). Scene ~ 90 km across.

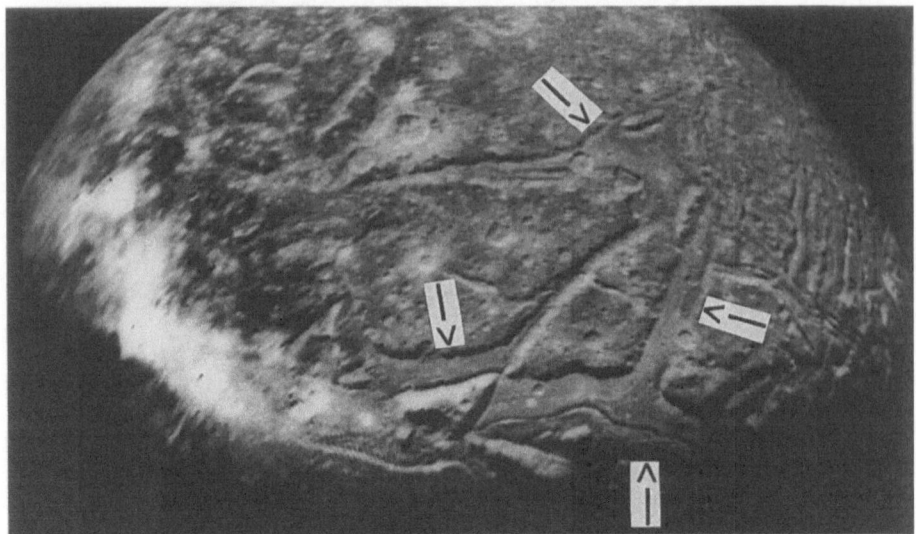

Figure 5. Image showing a complex system of canyons and one of several prominent, viscous flows on Ariel, some of which are over 1 km thick. The largest flows occur on the troughs' floors (arrows).

Figure 6. Ruach Planitia, Triton, and vicinity. The planitia is about 200 km across, and includes a scalloped wall ~200 m high, an inner terrace, a mostly smooth floor, an central area of complex pits and hummocks, and a sinuous rille. This feature resembles volcanic calderas and lava lakes on Earth, Venus, Mars, and Io. The region outside and to the left of the planitia appears to consist of multiple flows that spilled from Ruach Planitia and neighboring Tuonela Planitia, part of which is visible at the top of the image.

Figure 7. Cipango Planum, Triton. This area may be the distal region of flows spilled from Ruach and Tuonela Planitiae. A probable thick flow margin, left of center, appears to have wrapped around a protruding hill. Right of center is a volcanic rift, along which several volcanic craters are aligned and from which smooth deposits appear to have issued. Just below and left of center is an elongate volcanic crater containing what appears to be a viscous domical effusion. Scene roughly 300 km across.

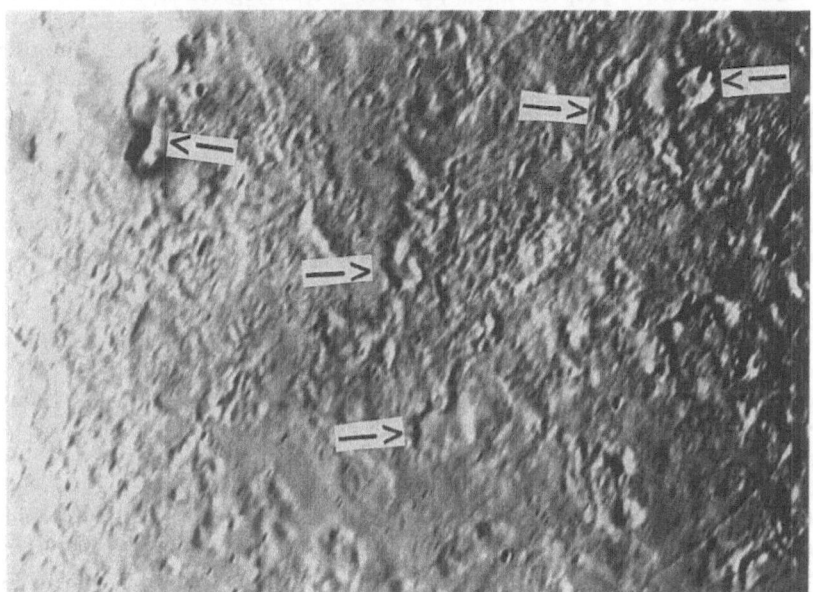

Figure 8. Area of southern Monad Regio, Triton, showing several volcanic craters and possible sinuous rilles oriented consistently from volcanic craters (arrows) toward the top of the image.

III. Properties and compositions of icy satellites

Remote reflectance spectroscopy has yielded a reasonably good knowledge of the major icy mineralogic components in the visible surface layers of all major icy satellites, a rough idea of the fraction and possible composition of nonice (rocky) components in this layer, and clues to the possible presence of certain minor components (*e.g.*, Clark *et al.* 1986). Almost all icy satellites have surfaces that are dominated by the spectral reflectance of water ice; the ice generally is intermixed with variable quantities of nonice components of low visual albedo and spectrally gray, yellowish, or reddish coloration. The nonice components may include rocky material and the photolysis products of carbonaceous ices. There are exceptions to these generalizations; Triton has a surface dominated by N_2 and other nonpolar molecules, and Enceladus' surface contains virtually none of the low-albedo components. In all cases, these data apply only to the visible surface microlayer. This layer may be nonrepresentative of the satellite as a whole or even of the material a meter below the surface because of the effects of igneous and gravitational differentiation and chemical processing in the space environment (especially electromagnetic and charged particle radiation).

Two other approaches to the possible compositions of icy satellites, summarized in Table 1, are based on (1) theoretical modeling of low-temperature nebular condensation of rocky and icy phases (*e.g.*, Lewis 1972, Prinn and Fegley 1988), and (2) observations of the compositions of comets and certain meteorites, which may represent the building blocks of some icy satellites. Cometary volatiles are particularly probable constituents of the satellites of Uranus and Neptune, because it is thought that comets mainly had their origins in the solar nebula near the orbits and Uranus and Neptune. Six model compositions are listed in Table 1 and show some of the possible variety of volatile assemblages and the types of liquids that could be generated from them. Some distinctive properties and possible volcanological behavior of a few of these liquids are summarized in Table 2. The predicted cryovolcanic liquids are of two general types-- aqueous solutions and non-polar molecular solutions. The possible planetary applications of these various lava types (Table 2) are highly speculative, but this listing highlights the probable chemical variety of cryovolcanism.

Pre-Voyager thermal models suggested that aqueous igneous differentiation may have been driven by radiogenic heating in icy satellites as small as 500 km in radius (Lewis 1971, Consolmagno and Lewis 1978). Besides the discovery that cryovolcanism affected satellites of half the minimum predicted size, it was unanticipated that the geomorphic expression of cryovolcanism could vary so tremendously from one object to another.

The small sizes of some partly melted satellites has been explained by (1) the highly efficient generation of heat in special cases by tidal dissipation (in addition to radiogenic and accretional heating; Ellsworth and Schubert 1983, Squyres *et al.* 1988), (2) a reduction in the melting point of multicomponent volatile assemblages relative to the melting point of pure ice (e.g., ice m.p. 273 K, ammonia dihydrate m.p. 176 K, ammonia-water-methanol m.p. ~153 K), and (3) conductive heat loss reduced by the insulating effects of substances of very low thermal conductivity (e.g., gas clathrate hydrates and porous megaregoliths). The geomorphic variety of cryovolcanism is partly explained by differences in the composition of the volatile-rich assemblages that constitute icy satellites (hence, differences in the composition and physical properties of partial melts). The possible importance of ammonia in lowering the melting point of ice (thereby making melting easier to accomplish on a small energy budget) was

TABLE 1. Some volatile condensate assemblages and their partial melts

Assemblage in rough order (to right) of decreasing temperature and degree of low-temperature chemical equilibration

Molecule	CI- or CM-chondrites + water ice, T_{cond} =160 K	NH$_3$-CH$_4$-rich nebula, homogeneous accretion near T_{cond} = 100 K (Lewis 1972)	NH$_3$-CH$_4$-rich nebula, homogeneous accretion near T_{cond} = 40 K (Lewis 1972)	NH$_3$-CH$_4$-rich nebula, heterogeneous accretion near T_{cond} = 40 K (Lewis 1972)	CO-N$_2$-rich nebula, T_{cond} = 40 K (Prinn and Fegley 1988)	Comets
H$_2$O	100	100	100	100	100	100
MgSO$_4$	0.5-2.5	0	0	0	0	0
NH$_3$	0	12	12[1]	12[3]	0.01	0.1-2.0
CH$_4$	0	0	60[2]	60[4]	< 12	0.7-2
H$_2$S	0	0	0	2.6[5]		
CO	0	0	0	0	100	1-7
CO$_2$	0	0	0	0	1	1.5-3.5
HCN	0	0	0.01	0		0.1-0.3
CH$_3$OH	0	0	0	0		0.1-10
H$_2$CO	0	0	0	0		1-10
N$_2$	0	0	0	0	12	1-10
Possible compositions of partial melts (melting temperature increases downward in list)	a. H$_2$O-MgSO$_4$ b. H$_2$O	a. H$_2$O-NH$_3$ b. H$_2$O	a. CH$_4$ b. H$_2$O-NH$_3$-CH$_4$ c. CH$_4$ and H$_2$O-CH$_4$ d. H$_2$O	a. CH$_4$ b. H$_2$O-NH$_3$-(NH$_4$)$_2$S-CH$_4$ c. NH$_3$-H$_2$O-(NH$_4$)$_2$S-CH$_4$ d. CH$_4$ and H$_2$O-CH$_4$ e. H$_2$O	a. N$_2$ b. CO-CH$_4$-N$_2$-CO$_2$ c. H$_2$O-CO$_2$-CH$_4$-N$_2$-CO d. H$_2$O	a. N$_2$ b. N$_2$-CO-CH$_4$-CO$_2$ c. H$_2$O-CH$_3$OH d. H$_2$O

T_{cond} is the condensation temperature.

[1] Other water-soluble salts, especially Na$_2$SO$_4$, are also abundant in CI and CM carbonaceous chondrites and should constitute significant potential solutes in cryovolcanic brines. Some of these salts or their reaction products with ammonia (e.g., (NH$_4$)$_2$SO$_4$) may also occur in melts produced in the other low-temperature condensate assemblages.

[2] Partly as clathrate hydrate, excess as pure methane ice.

[3] Partly as sulfide.

[4] As pure methane ice.

[5] As ammonium sulfide.

recognized prior to Voyager. It was not until fairly late in the Voyager mission that the possible importance of ammonia and other volatiles besides water was recognized with regard to (1) other physical properties of cryovolcanic liquids, such as viscosity, density, and vapor pressure, and (2) cryovolcanic landform development. The Voyager 2 Uranus encounter and then the Neptune/Triton flyby revealed types of cryovolcanic features, especially thick effusive flows and numerous features formed by explosive

TABLE 2. Properties of selected candidate cryomagmas (eutectoid compositions)

Liquid	Liquid composition, mass %	Melting point, K	Liquid density, g cm^{-3}	Viscosity of liquid, poises	Solid composition, mass %	Solid density, g cm^{-3}	Some possible planetary appplications
Water	H_2O 100%	273	1.000	0.017	Ice 100%	0.917	Plains volcanism on Europa and Ganymede
Brine	H_2O 81.2% $MgSO_4$ 16% Na_2SO_4 2.8%	268	1.19	0.07	Ice 50% $MgSO_4.12H_2O$ 44% $Na_2SO_4.10H_2O$ 6%	1.13	Plains volcanism on Europa and Ganymede
Ammonia-water	H_2O 67.4% NH_3 32.6%	176	0.946	40	$NH_3.2H_2O$ 97% $NH_3.1H_2O$ 3%	0.962	Plains volcanism on Dione, Tethys, and Enceladus
Ammonia-water-non-polar gas	H_2O 67% NH_3 33% CH_4 0.1-2%	176	0.94	40	$NH_3.2H_2O$ 97% $NH_3.1H_2O$ 3%	0.96	Explosive volcanism and cinder cones on Triton
Ammonia-water-methanol	H_2O ~47% NH_3 ~23% CH_3OH ~30%	~153	~0.978	~40,000	$NH_3.1H_2O$ ~46% $CH_3OH.1H_2O$ ~54%		Thick flows on Triton, Ariel, and Miranda
Nitrogen-methane	N_2 86.5% CH_4 13.5%	62	0.783	0.003	100% N_2-CH_4 solid sol'n @ 62 K; 88% N_2-CH_4 + 12% CH_4-N_2 @38 K		Sapping fluid, sublimable lava, and "geyser" gas on Triton

("cryoclastic") volcanism, that were unobserved on the satellites of Jupiter and Saturn. It seemed that these features would be difficult to explain by eruptions of pure water, so that attention has since focussed on the chemically more complex aqueous solutions of ammonia and other volatiles (Stevenson 1982, Squyres et al. 1983, Croft and Soderblom 1991, Kargel et al. 1991, Schenk 1991, Kargel 1992, Croft et al. 1995). Alternatively, some of the thicker flows on the Uranian satellites have been explained by eruptions of warm, solid water ice or ice that was somewhat softened by inclusion of small amounts of methane or other volatiles (Jankowski and Squyres 1988).

Although knowledge of the compositions of icy satellites and cryovolcanic deposits is rather poor, compositional variations of the first order are apparent from the densities of icy satellites, which indicate large differences in rock:ice ratios (Johnson *et al.*, 1987). "Rock" may include silicate, metal sulfides, metallic iron-nickel, graphite, and other relatively dense and involatile phases, and "ices" may include water ice, ammonia hydrate, methane clathrate hydrate, nitrogen ice, and other highly volatile and low-density molecular ices. The average densities of a few satellites are very precisely known and the densities of others are roughly known from spacecraft radio tracking and observations of mutual gravitational perturbations. Most of the observed densities and rock:ice ratios fall within the range predicted by theoretical models of nebular condensation under different sets of assumed or modeled conditions. For example, condensation in a $CO-N_2$-rich nebula results in a larger rock:ice ratio than condensation in a CH_4-NH_3-rich nebula (Johnson *et al.* 1987, Prinn and Fegley 1988). Other processes can cause substantial volatile fractionations in the solar and circumplanetary nebulae, so that the variations in satellite densities or rock:ice ratios are unreliable indicators of nebular processes, conditions, and volatile assemblages; one may merely construct reasonable but nonunique models that fit the constraints. Regardless of exactly how the different rock:ice ratios arose, an important implication is that compositional variations in the volatile contents of icy satellites are also likely. Satellite densities are even less helpful with the volatile composition, because most of the predicted volatile ices have roughly comparable densities and cause similar effects on the bulk density of any mixture of rock and ices (Lewis 1972, Prinn and Fegley 1988).

IV. Possible compositions and properties of cryolavas

Ubiquitous water ice features in the reflection spectra of all major icy satellites, except Triton, suggests that cryovolcanic flows are aqueous (Clark *et al.* 1986). The flows on Triton, too, are believed to be aqueous but covered by frosts of nonaqueous substances (mainly nitrogen with traces of other volatiles, Cruikshank *et al.* 1991, Croft *et al.* 1995). Triton is so distant from the sun, its surface so cold, and its mass so great that outgassed nitrogen and other molecular ices are retained as surfaces ices and in a tenuous atmosphere. However, Triton is warm enough and nitrogen and methane are volatile enough that any original volcanic flows of these substances probably would have long ago sublimed. Any substantial relief formed in frozen nitrogen by volcanic or tectonic activity would not only tend to sublimate but it would also tend to flow glacier-style very rapidly (even at Triton's frigid 38 K) until sharp local relief disappeared (Croft *et al.* 1995). Hence, it is thought that Triton's substantial surface relief is formed in aqueous substances that are covered by nitrogen-rich frosts (Croft *et al.* 1994).

Titan possesses a dense atmosphere dominated by nonpolar gases (mainly nitrogen plus some methane, heavier hydrocarbons, and probably argon), although these gases cannot form stable solids on the surface (condensed liquid solutions may occur on Titan). The other icy satellites of Jupiter, Saturn, and Uranus are generally too warm for solid nitrogen and methane ices to exist, and these objects are generally too small to retain a substantial atmosphere; hence, most outgassed or extruded nitrogen and methane on the majority of icy satellites has probably been lost to space. Large amounts of gas clathrate hydrates may still be trapped in the satellites' interiors.

In sum, it is generally thought that cryovolcanic landscapes throughout the outer solar system are fundamentally aqueous. The general analogy between certain cryovolcanic terrains, especially those on Triton, and silicate volcanic terrains on the rocky planets raises the intruiguing question of how aqueous solutions could have behaved much like silicate lavas so as to produce similar landforms. Multicomponent phase equilibria in the types of chemical systems indicated in Table 2 have helped in the identification of some likely aqueous lavas (Kargel 1991, 1992). Other laboratory measurements have shown that the rheological characteristics of solutions in the water-ammonia-methanol series (a possible class of cryovolcanic lava) are comparable to the series of silicate lavas ranging from basalt to dacite (Kargel *et al.* 1991).

Laboratory studies have highlighted the volcanological importance of what may be minor constituents in bulk satellites but major constituents in multicomponent aqueous lavas. It is noteworthy that minor constituents in the Earth, such as Na, K, and H_2O, have petrologic and geologic importance disproportionate to their abundances (0.25%, 0.02%, and ~0.03% of Earth's mass, respectively). For instance, these components are essential ingredients in the formation of granitoid rocks, and, thus, are keys to the origin of Earth's continental crust, to say nothing of the hydrosphere and biosphere. The importance of Na, K, and H_2O to Earth's structure and geology is magnified because of the lithophile incompatible chemical nature of these components (i.e., they are concentrated in partial melts, and, thus, in Earth's crust). Likewise, minor water-soluble components of icy satellites, probably including many substances not mentioned in Tables 1-2, could be key ingredients in determining the many unique geologic histories, interior structures, and surface morphologies of icy satellites.

The space-filling characteristic of the plains on the Jovian and Saturnian satellites suggests that the flow substance(s) had a fairly low viscosity, perhaps like that of basaltic lava or something even less viscous (such as water, brine, or ammonia-water lacking large amounts of other chemical constituents or suspended crystals). By contrast, the cryovolcanic flows on the satellites of Uranus and Neptune are very thick and generally form lobate deposits and, on Triton, discrete volcanic constructs. This morphologic evidence suggests a profound rheological difference between the flows on the satellites of Jupiter/Saturn and those on the satellites of Uranus/Neptune. A simple accounting for this difference is that chemical variations caused flows to vary in viscosity (for example, due to different types of molecular bonding or different melting temperatures). The inferred rheological characteristics and the distribution of these two types of flows can be explained by the following working hypothesis.

(1) Flows on the Jovian and Saturnian satellites are thought to contain water and perhaps ammonia and salts, but not large quantities of other water-soluble substances (such as methanol) that would substantially further increase the liquids' viscosities.

(2) Flows on the Uranian and Neptunian satellites are thought to include, besides water, ammonia, and salts, large amounts of additional substances (such as methanol and/or hydrogen sulfide) that are highly soluble in low-temperature aqueous solutions and that would greatly increase the liquids' viscosities (Table 2) (Kargel *et al.* 1991).

These differences are consistent with the inclusion of a large amount of cometary or interstellar ices in the Uranian and Neptunian satellites, and the absence of a large fraction of such ices in the Jovian and Saturnian satellites. Cometary or interstellar components would introduce large amounts of clathrate hydrates, which would yield high-pressure aqueous liquids saturated in nonpolar gases (up to several weight percent); the gas-charged liquid would tend to devolatilize and cause explosive volcanism, consistent with extensive explosive cryovolcanism on Triton (Kargel and Strom 1990). An utter lack of any indication of explosive volcanism on Ganymede and Europa is strong evidence that nonpolar gases (in the form of clathrate hydrates) are virtually absent from the interior zones of these objects where partial melting occurred. Possible explosive volcanism on Enceladus sugests that it may be compositionally transitional somewhere between the satellites of Jupiter and those of Uranus and Neptune.

V. References

Burns, J.A. (1986) Some background about satellites, in J.A. Burns and M.S. Matthews (eds.), *Satellites*, University of Arizona Press, Tucson, pp. 1-38.

Clark, R.N., Fanale, F.P., and Gaffey, M.J. (1986) Surface composition of natural satellites, in J.A. Burns and M.S. Matthews (eds.), *Satellites*, University of Arizona Press, Tucson, pp. 437-491.

Consolmagno, G.J., and Lewis, J.S. (1978) The evolution of satellite interiors and surfaces, *Icarus* **34**, 280-293.

Croft, S.K. (1992), Proteus: Geology, shape, and catastrophic disruption, *Icarus* **99**, 402-419.

Croft, S.K., and Soderblom, L.A. (1991) Geology of the Uranian satellites, in J.T. Bergstralh, E.D. Miner, and M.S. Matthews (eds.), *Uranus*, University of Arizona Press, Tucson, pp. 561-628.

Croft, S.K., Kargel, J.S., Kirk, R.L., Moore, J.M., Schenk, P.M., and Strom, R.G. (1995) The geology of Triton, in D.P. Cruikshank (ed.), *Neptune and Triton*, University of Arizona Press, Tucson (in press).

Cruikshank, D.P., *et al.* (1991) Tentative identification of CO and CO_2 ices on Triton, *Bull. Amer. Astronom. Soc.* **23**, 1208.

Ellsworth, K., and G. Schubert (1983) Saturn's icy satellites: Thermal and structural models, *Icarus* **54**, 490-510.

Farinella, P., Paolicchi, P., Strom, R.G., Kargel, J.S., and Zappalá, V. (1990) The fate of Hyperion's fragments, *Icarus* **83**, 186-204. .

Jankowski, D.G., and S.W. Squyres (1988) Solid-state ice volcanism on the satellites of Uranus, *Science* **241**, 1322.

Johnson, T.V., R.H. Brown, and J.B. Pollack (1987) Uranus satellites: Densities and composition, *J. Geophys. Res.* **92**, 14,884-14,894.

Kargel, J.S., and R.G. Strom (1990) Cryovolcanism on Triton, *Lunar Planet. Sci. Conf.* **XXI**, 599-600.

Kargel, J.S., 1991, Brine volcanism and the interior structures of asteroids and icy satellites, *Icarus* **94**, 368-390.

Kargel, J.S., S.K. Croft, J.I. Lunine, and J.S. Lewis (1991) Rheological properties of ammonia-water liquids and crystal-liquid slurries: Planetological applications, *Icarus* **89**, 93-112.

Kargel, J.S. (1992) Ammonia-water volcanism on icy satellites: Phase relations at 1 atmosphere, *Icarus* **100**, 556-574.

Kirk, R.L., Soderblom, L.A., Brown, R.H., Kieffer, S.W., and Kargel, J.K. (1995) Triton's plumes: Discovery, characteristics, and models, in D.P. Cruikshank (ed.), *Neptune and Triton*, The University of Arizona Press, Tucson (in press).

Lewis, J.S. (1971) Satellites of the outer planets: Their chemical and physical nature, *Icarus* **15**, 174-185.

Lewis, J.S. (1972) Low-temperature condensation from the solar nebula, *Icarus* **16**, 241-252.

Lucchitta, B.K., and Soderblom, L.A. (1982) The geology of Europa, in D. Morrison (ed.), *Satellites of Jupiter*, The University of Arizona Press, Tucson, pp. 521-555.

Prinn, R.G., and B. Fegley, Jr. (1988) Solar nebula chemistry: Origin of planetary, satellite, and cometary volatiles, in S.K. Atreya, J.B. Pollack, and M.S. Matthews (eds.), *Planetary and Satellite Atmospheres: Origin and Evolution*, University of Arizona Press, Tucson, pp. 78-136.

Rothery, D.A. (1992) *Satellites of the Outer Planets: Worlds in Their Own Right*, Oxford University Press, New York, pp. 208.

Schenk, P.M. (1990) Fluid volcanism on Ariel and Miranda: Flow morphology and composition, *J. Geophys. Res.* **96**, 1887-1906.

Shoemaker, E.M., Lucchitta, B.K., Plescia, J.B., Squyres, S.W., and Wilhelms, D.E. (1982), The geology of Ganymede, in D. Morrison (ed.), *Satellites of Jupiter*, The University of Arizona Press, Tucson, pp. 435-520.

Smith, B.A., *et al.* (1979) The Jupiter system through the eyes of Voyager 1, *Science* **204**, 951-972.

Smith, B.A., *et al.* (1982) A new look at the Saturn system: The Voyager 2 images, *Science* **215**, 504-536.

Smith, B.A., *et al.* (1986) Voyager 2 in the Uranian system: Imaging science results, *Science* **233**, 43-64.

Smith, B.A., *et al.* (1989) Voyager 2 at Neptune: Imaging science results, *Science* **246**, 1422-1429.

Squyres, S.W., R.T. Reynolds, P.M. Cassen, and S.J. Peale (1983) The evolution of Enceladus, *Icarus* **53**, 319-331.

Squyres, S.W., R.T. Reynolds, A.L. Summers, and F. Shung (1988) Accretional heating of the satellites of Saturn and Uranus, *J. Geophys. Res.* **93**, 8779-8794.

Stevenson, D.J. (1982) Volcanism and igneous processes in small icy satellites, *Nature* **298**, p. 142-144.

Strom, R.G. (1986) The solar system cratering record: Voyager 2 results at Uranus and implications for the origin of impacting objects, *Icarus* **70**, 517-535.

Thomas, P.C. (1989) The shapes of small satellites, *Icarus* **77**, 248-274.

FORMATION OF SATELLITE AND RING SYSTEMS : COMPARATVE ASPECTS

D. MÖHLMANN
DLR-Institut für Raumsimulation
51140, Köln, Germany

Abstract. There are four systems of a massive central body with a regularily structured satellite system in the Solar system: the planetary system and the satellite systems of Jupiter, Saturn and Uranus. Comparable structures in these four systems can be understood as indications for comparable processes of origin and formation. It is the aim of this paper to describe comparable properties, and to discuss possible physical processes in pre-satellite disks which can be the cause for this comparability.

1. The Four Systems

Distances from the central body, orbital inclinations and radii of the satellites are given with Tables 1 - 4 for the satellite systems of the Sun, Jupiter, Saturn and Uranus (the data for Tables 1-5 were taken from K.R.Lang, 1992).

It is interesting to note that there are comparable subgroups in these systems. There is an inner group of small bodies at low inclination orbits, a group of the "principal satellites" (described by Table 5), containing most of the mass of the satellite system, and an outer group of smaller bodies at high inclination (including retrograde) orbits. The outer group may consist of captured (external) bodies and of bodies, belonging originally to the system and being scattered gravitationally outward from the more massive inner regions. The satellites of the inner group may have formed there, or they have been scattered gravitationally inward from the more massive region of the principal satellites.

As an additional clue to understand processes in the early solar nebula also some of the observed structures in planetary rings have to be taken into account (Lissauer, Cuzzi,1985).

Throughout this paper, the principal satellites and their comparable properties shall be discussed mainly. They contain most of the mass of the satellite systems, and their formation was the key process in the early evolution of these systems.

Earth, Moon, and Planets **67**: 115–129, 1995.

TABLE 1

Jovian Satellite System

Satellite	Distance from Planet Center (10^6 m)	Inclination (degrees)	Radius (10^3m)
J14 Adrastea	128	[a]0	20±5
J16 Metis	128	[a]0	20±5
J5 Amalthea	181	0.4	135×85×75
J15 Thebe	221	[a]0	40±5
J1 Io	422	0.0	1815
J2 Europa	671	0.5	1569
J3 Ganymede	1 070	0.2	2631
J4 Callisto	1 880	0.2	2400
J13 Leda	11 110	26.7	[a] 5
J6 Himalia	11 470	27.6	90±10
J10 Lysithea	11 710	29.0	[a]10
J7 Elara	11 740	24.8	40±5
J12 Ananke	20 700	147	[a]10
J11 Carme	22 350	164	[a]15
J8 Pasiphae	23 300	145	[a]20
J9 Sinope	23 700	153	[a]15

TABLE 2

Saturnian Satellite System

Satellite	Distance from Planet Center (10^6m)	Inclination (degrees)	Radius (10^3m)
S17 Atlas	137.7	[a]0	20×10
S16 Prometheus	139.4	[a]0	70×50×40
S15 Pandora	141.7	[a]0	55×45×35
S10 Janus	151.4	[a]0	110×100×80
S11 Epimetheus	151.5	[a]0	70×60×50
S1 Mimas	186	1.5	196
S2 Enceladus	238	0.0	250
S3 Tethys	295	1.1	530
S13 Telesto	295		17×14×13
S14 Calypso	295		17×11×11
S4 Dione	377	0.0	560
S12 Helena	377	0.2	18×16×15
S5 Rhea	527	0.4	765
S6 Titan	1 222	0.3	2 575
S7 Hyperion	1 481	0.4	205¥130¥110
S8 Iapetus	3 561	14.7	730
S9 Phoebe	12 954	150	110±10

TABLE 3
Uranian Satellite System

Satellite	Distance from Planet Center (10^6m)	Radius (10^3m)
U13 Cordelia	49.7	[a]20
U14 Ophelia	53.8	[a]25
U15 Bianca	59.2	[a]25
U9 Cressida	61.8	[a]30
U12 Desdemona	62.7	[a]40
U8 Juliet	64.6	[a]40
U7 Portia	66.1	[a]40
U10 Rosalind	69.9	[a]30
U11 Belinda	75.3	[a]30
U6 Puck	86.0	85±5
U5 Miranda	129.9	236
U1 Ariel	190.9	579
U2 Umbriel	266.0	586
U3 Titania	436.3	790
U4 Oberon	583.4	762

Note that only Miranda has a notable inclination of 3.4°.

TABLE 4
The Solar System

Planet	Distance from the Sun (10^{11}m)	Inclination	Equatorial Radius (10^6m)
Mercury	0.579	7.0	2.439
Venus	1.082	3.39	6.051
Earth	1.496	0	6.378
Mars	2.279	1.85	3.397
Jupiter	7.783	1.31	71.492
Saturn	14.270	2.49	60.268
Uranus	28.696	0.77	25.559
Neptune	44.966	1.77	24.764
Pluto/Charon	59.00	17.15	1.123/0.56

Note that the satellites of Neptune were not taken into account. This system seems to have been modified essentially toward non-regular structures by processes, related to the probable capture of Triton.

2. Angular Momenta and Radial Extension

There is a remarkable discrepancy between the above mentioned three planetary satellite systems and the planetary system in the distribution of angular momenta beween the spin of the central body and the orbital angular momentum of the satellite systems.

While most of the angular momentum of the Solar system is in the orbital motion of the giant planets, the spin of the central mass dominates the angular momenta in the three planetary satellite systems.

Table 6 describes this more in detail. This difference in dynamic properties is sometimes interpreted as an indication for different origin and formation of these systems, and as a hint for a Laplace-type formation of the planetary

TABLE 5
Principal Satellites

Satellite	Distance $(10^6 m)$	Radius $(10^3 m)$	Mass $(10^{20} kg)$	Density $(10^3 kg/m^3)$
		Jupiter		
J1 Io	422	1 815	892	3.55
J2 Europa	671	1 569	487	3.04
J3 Ganymede	1 070	2 631	1 490	1.93
J4 Callisto	1 880	2 400	1 075	1.83
		Saturn		
S1 Mimas	186	196	0.455	1.44
S2 Enceladus	238	250	0.74	1.13
S3 Tethys	295	530	7.55	1.20
S4 Dione	377	560	10.52	1.41
S5 Rhea	527	765	24.9	1.33
S6 Titan	1 222	2 575	1 346	1.88
(Iapetus	3 561	730	18.8	1.15)
		Uranus		
U5 Miranda	129.9	236	0.8	1.25
U1 Ariel	190.9	579	13.5	1.55
U2 Umbriel	266.0	586	12.7	1.58
U3 Titania	436.3	790	34.8	1.69
U4 Oberon	583.4	762	29.2	1.64

satellite systems from that mass, which was left behind from the rotationally unstable forming central bodies.

That this is not necessarily the case can be seen from the radially restricted "sphere of influence" of these planets. In other words, these satellite systems are restricted in their radial extension by the disturbing influence of the gravitation of the Sun. The scales of these limitations in extension of the planetary satellite systems are given in Table 7, indicating that bodies in the outer subgroup of satellites may be influenced over longer timescales by "disturbing" solar gravitation.

TABLE 6
Angular momenta

	Orbital momentum L of the satellite system (kg m^2/s)	Spin S of the central mass (kg m^2/s)	ratio L/S
Jupiter	$4.50 \; 10^{36}$	$6.72 \; 10^{38}$	$6.7 \; 10^{-3}$
Saturn	$9.57 \; 10^{35}$	$8.72 \; 10^{37}$	$1.1 \; 10^{-2}$
Uranus	$1.40 \; 10^{34}$	$2.0 \; 10^{36}$	$7 \; 10^{-3}$
Sun	$3.15 \; 10^{43}$	$1.7 \; 10^{41}$	185.3

Consequently, the orbital angular momentum of the planetary satellites, which is proportional to the square root of the orbital radius, is restricted by the limited extension of stable regions around the planets. This constraint did not exist during the formation of the Solar system. The difference in angular momenta distribution is therefore not necessarily connected with a different origin of these systems.

TABLE 7
Spheres of Influence

$$\text{Hill sphere} \quad r_H = R_{orb}\left(\frac{m_{Pl}}{M_{Sun}}\right)^{1/3}$$

Sphere of gravitational equilibrium

$$r_E = R_{orb}\left(\frac{m_{Pl}}{M_{Sun}}\right)^{1/2}$$

Planet	r_H (m)	r_E (m)
Mercury	3.19×10^8	2.37×10^7
Venus	1.46×10^9	1.69×10^8
Earth	2.16×10^9	2.59×10^8
Mars	1.56×10^9	1.09×10^8
Jupiter	7.66×10^{10}	2.40×10^{10}
Saturn	9.42×10^{10}	2.42×10^{10}
Uranus	1.01×10^{11}	1.90×10^{10}
Neptune	1.67×10^{11}	3.22×10^{10}

3. Vertical Disk Structure

The equilibrium in a circumstellar or circum-protoplanetary disk is governed by thermal pressure of the disk and gravitation of the central mass. For the vertical "z-"components this can be described by

$$\frac{1}{\rho}\frac{dp}{dz} = -\gamma\frac{M_c}{r^2}\frac{z}{r} \tag{1}$$

where M_c is the central mass, r the radial distance from the central body, and p and r are pressure and mass density, respectively. They can be related by an equation of state $p=\rho c^2$ with "c" as a velocity of sound.

The solution is

$$\rho = \rho_0 \exp\{-\frac{z^2}{2H^2}\} \qquad (2)$$

where ρ_0 is the midplane mass density. The "scale height" H is given by

$$H = \frac{c}{\Omega_K} \qquad (3)$$

and Ω_K is the Keplerian angular velocity

$$\Omega_K = \gamma \frac{M_c}{r^3} \qquad (4)$$

with the constant of gravitation γ. Consequently, the ratio of scale height and radial distance is given by

$$\frac{H}{r} = \frac{c}{\Omega_K r} = \frac{c}{v_\varphi} \qquad (5)$$

where ϖ_φ is the Keplerian azimutal orbital velocity.

Fig. 1 gives this ratio for the four different systems under consideration.

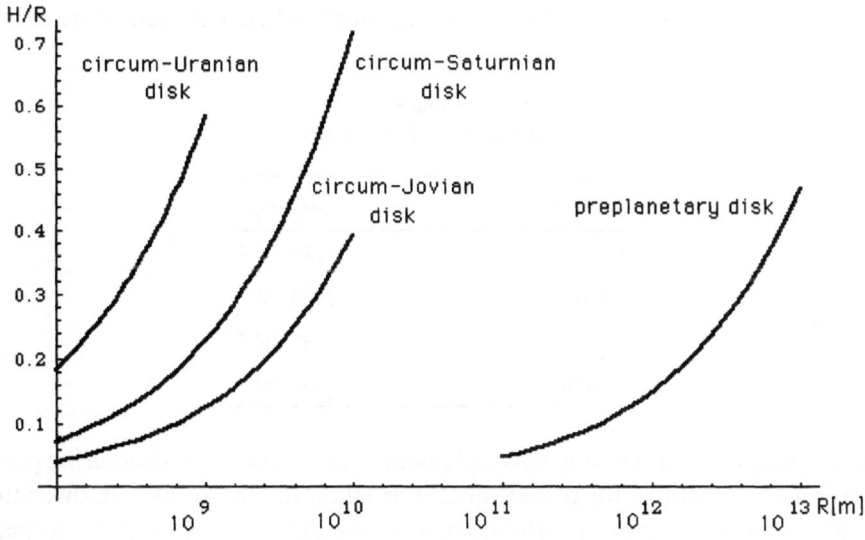

Fig. 1 Relative disk scale heights

Obviously, disk scale heights are comparable in these systems, they are of the order of a few tenths of the radial distance. For further computations, a value of H=0.1 r will be used throughout this paper. The corresponding volume of a disk is then $V=\pi R^2 H \cong 0.3\ R^3$, where R is the radial extension of the disk.

4. Gravitational Stability of Disks

It has been discussed intensively in the literature (Cameron,1978) that the self-gravitation of a preplanetary or pre-satellite disk might be able to cause local gravitational instabilities which might be trigger mechanisms for the growth of solid bodies in the km- to 10 km scale, called "planetesimals". It can be shown from corresponding dispersion relations, derived by linearized perturbation theory, that a condition for self-gravitation to overcome the disrupting action of the Keplerian shear motion, described by the Keplerian angular velocity, is given by

$$\omega_g^2 >> \Omega_K^2 \qquad (6)$$

where the frequency for effects due to disk self-gravitation is given by $\omega_g^2 = 4\pi\gamma\rho$ with the disk mass density ρ. With a disk mass m_d, a density $\rho=m_d/V$ and the above given disk volume $V=0.3\ R^3$ it follows as a necessary condition for self-gravitation to overcome disrupting effects of the gravitation of the central body that

$$\frac{m_d}{M_c} \ge \frac{1}{40} \qquad (7)$$

It can be seen from Table 8 that this condition is not fulfilled in the four systems under consideration.

Table 8
Disk/Central body mass ratio

System	m_d/M_c
Sun	$1.34\ 10^{-3}$
Jupiter	$2.08\ 10^{-4}$
Saturn	$2.48\ 10^{-4}$
Uranus	$1.05\ 10^{-4}$

Consequently, self-gravitation was not essential as a large scale structuring process or a trigger mechanism for planetesimal formation in the presatellite disks under consideration, or, most of the original mass of the disk was lost after the formation of solid bodies and satellites.

5. Tidal Forces

Tidal forces can be essential for the evolution of disks. It can be seen from Table 9 that these forces can have been much more effective in the small presatellte disks than in the planetary system. The consequences of the action of tidal forces are at one side the increased potential for destruction of local structures in the disk, and on the other side, an increased outward angular momentum transport. These processes can be essential for the evolution of a disk above a central mass. They can be a cause for the different properties of the preplanetary and the presatellite disks.

Table 9
Tidal forces

$$f = -\gamma\frac{m\,M_c}{r^2} + \Omega_K^2\,r = -\gamma\frac{mM_c}{r^2} + \gamma\frac{mM_c}{R^3}\,r$$

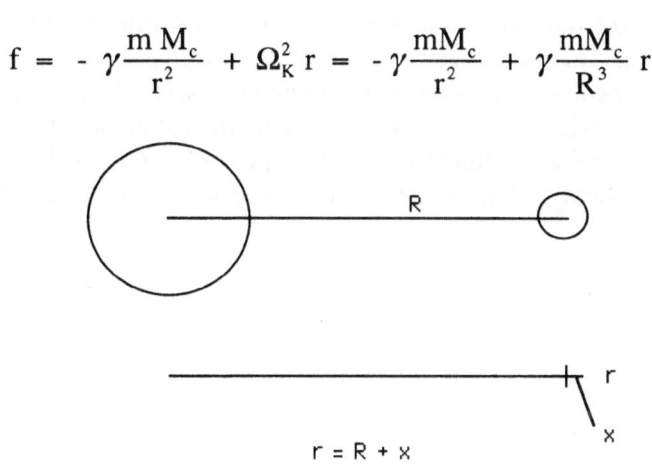

$$r = R + x$$

$$f(x) = 2\gamma\frac{mM_c}{R^3}\,x + \gamma\frac{mM_c}{R^3}\,x = 3\,\gamma\frac{mM_c}{R^3}\,x$$

System	$3gM_c/R^3$
Sun	$1.18\ 10^{-13}$ (Earth)
Jupiter	$3.80\ 10^{-10}$ (10^9m)
Saturn	$1.14\ 10^{-10}$ (10^9m)
Uranus	$1.40\ 10^{-10}$ ($5\ 10^8$m)

6. Compositional Gradients

The composition of the planets in the Solar systems follows a clear radial gradient with refractory matter of higher density in the inner parts and volatiles and condensates in the outer parts of the system. The inner planets are stony with a high iron content, while the giant planets are gas-dominated and the outer planets seem to "ice-planets". This "composition-gradient" is related to the temperature in the preplanetary disk, and insofar, a similar picture can be estimated to exist in the planetary satellite systems. Table 10 describes the composition in these systems.

A trend, similar to that in the planetary system can be found again in the Jovian system, but not in the others, indicating also for the Jovian system a high temperature inner disk. This leads to a very interesting conclusion and question. Are the satellites of the inner group of the Jovian satellites made of refractories? This should be assumed in analogy to the terrestrial planets. A positive answer would be a verification of the theoretical approach of hot inner disks. On the other hand, there is a further question. Why is there no compositional gradient in the satellite systems of Saturn and Uranus? Was the luminosity of these protoplanets not sufficient to heat up the disks sufficiently? This seems to be a challenge for future theoretical approaches.

Table 10
Collisional gradients

Satellite	Density $(10^3 kg/m^3)$	Compositional Features
Io	3.55	Rock
Europa	3.04	Rock+100km H_2O-layer
Ganymede	1.93	60% rock , 40% ice
Callisto	1.83	60% rock , 40% ice
Mimas	1.44	~40% rock, 60% ice
Enceladus	1.13	"
Tethys	1.20	"
Dione	1.41	"
Rhea	1.33	"
Titan	1.88	60% rock, 40% ice+N_2, CH_4
Miranda	1.25	icy ?
Ariel	1.55	icy ?
Umbriel	1.58	icy ?
Titania	1.69	Rock-ice ?
Oberon	1.64	Rock-ice ?

7. Disk Temperatures

Energy sources for preplanetary and pre-satellite disks are the friction caused dissipation in the Keplerian shear motion and the heating from the growing protosun or protoplanet. At the outer edge of the disk, energy is lost by radiation, as described by the Stefan-Boltzmann law with the radiation constant σ.

Neglecting any additional internal heating, caused by the opacity of disk matter, the disk temperature T can be estimated by equating the radiative energy loss by the dissipation generated heat, which is proportional to the kinematic viscosity ν :

$$\frac{9}{4}\nu\,\Omega_K^2 \;=\; \frac{2}{\Sigma}\sigma\,T^4 \tag{7}$$

where Σ is the surface mass density of the disk.

Estimating the viscosity from the friction caused inward motion v_r of the disk matter via $v_r=-3\nu/2r$ and using the equation of continuity in the form $2\pi r\Sigma(-v_r) = \dot{M}$, there follows

$$T^4 = (3\gamma \,/\, 8\pi\sigma)M\dot{M} \,/\, r^3 \tag{8}$$

where $\dot{M}$ is the growth rate of the central body. Note that additional heating by the luminosity of the central body and heating due to opacity were not taken into account. These should have been essential additional heat sources, at least in the case of the Solar system and the Jovian system. Values, following with $\dot{M}=M_c/10^6$years are given in Fig. 2 (with $T_e=T$). For more details, the reader is referred to Lin and Papaloizou (1985).

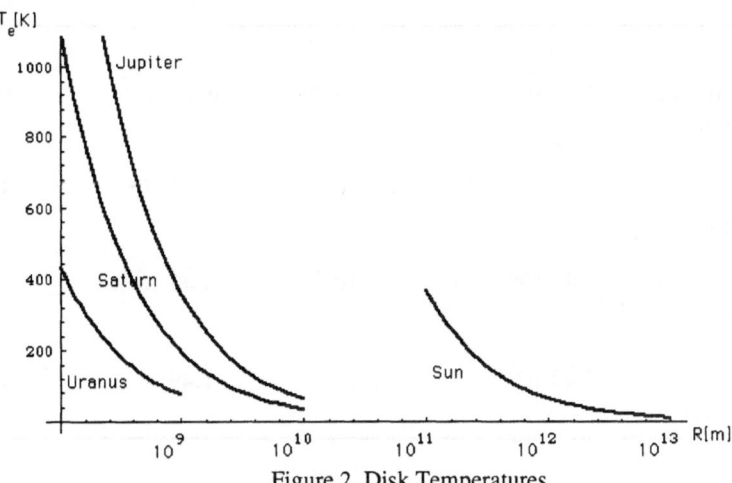

Figure 2. Disk Temperatures

It is an interesting result of Fig. 2 that the temperatures in the disks of the four systems under consideration are in comparable ranges.

8. Disk Surface Mass Densities

A key parameter to understand disks around massive cental bodies is the (height integrated) surface mass density. Using the value of $\zeta = m_{dust}/m_{gas} = 0.0034$, as it was proposed by Podolak and Cameron, there follows $m_{disk} \approx 300 m_{dust} \approx 300 m_{sat}$, where m_{sat} is the mass of the present satellite system. So, and in the sense of an estimation, the pre-satellite disk surface mass density can be related to known values via $\Sigma = m_{disk}/pr^2$. Table 11 gives the corresponding values for the four systems under consideration.

Here, the ·midplane central pressure is mentioned too, as it follows from $p = \Sigma \Omega_K c / 4$. It is interesting to note that the pressure in the pre-satellite disks is much higher than in the preplanetary disk. This might have been an essential difference between these disks (note that fluid phases seem to be possible under these conditions; This might be essential for growth processes). To estimate the relative importance of self-gravitation, as discussed in Chapter 4, the ratio of the two characteristic frequencies is given too with Table 11. It can be seen, that self-gravitation was too weak to be a dominating process.

Table 11
Physical parameters

System	Surface mass density [kg/m²]	Mass density [kg/m³]	Central pressure [mbar]	w_g^2/W_K^2
Sun (<4.5 10^{12}m) (at Jupiter)	1.25 10^4	8.03 10^{-8}	8.93 10^{-4}	0.24
Jupiter (Europa)	1.04 10^7	7.7 10^{-2}	7.32 10^2	0.15
Saturn (Rhea)	7.86 10^6	7.5 10^{-2}	4.43 10^2	0.24
Uranus (Umbriel)	2.42 10^6	4.5 10^{-2}	1.49 10^2	0.12

It can be seen from Table 11 that the mass densities in the pre-satellite systems were greater than in the preplanetary disk. But it has to be noted, that all these estimations are based on the model of a more or less stable disk. Very probably, these disks have to be seen also as a transient phenomenon, governed in its time scales by internal dissipation, solidification and clearing and by the inflow of fresh matter. Therefore, the timescales for the evolution of these disks have to be studied.

9. Sedimentation and Particle Growth

The frictional force between gas and a dust particle in a gas with a free mean path larger than the particle size is given by the Epstein law, describing the frictional acceleration as

$$a = \frac{\rho_g}{\rho_p} \frac{\bar{v}}{r_p} \Delta v \tag{9}$$

where ρ_p and ρ_g are the mass densities of the gas and the dust particle, $\bar{v}$ is the mean thermal speed of the gas particles, ρ_p is the radius of the dust particle and Δv is the velocity of the gas relative to the dust particle. This leads to an equation of (vertical) motion

$$\ddot{z} = -\frac{\rho_g}{\rho_p} \frac{\bar{v}}{r_p(t)} \dot{z} - \Omega_K^2 z \tag{10}$$

where the growth of the dust particle by colliding and sticking (sticking probability w_{stick}) can be described by

$$\frac{dr_p}{dt} = \frac{w_{stick}}{4} \frac{\rho_{dust}}{\rho_p} |\dot{z}| \tag{11}$$

This gives

$$r = r_0 + (1 - \frac{z}{z_0}) \frac{w_{stick}}{8} \frac{\Sigma}{\rho_p} \tag{12}$$

for a particle, starting with a radius $r(0)=r_0$ at the scale height $z(0)=z_0=H$. Here, $\Sigma=2H\rho$ has been used. It has been shown by Weidenschilling that dust particles can grow by collisional aggregation towards the 10cm to meter scales in size (Weidenschilling and Cuzzi, 1993), if not only vertical settling but also collisional growth during inward motions in the midplane are taken into account. But it has to be mentioned here that the sticking probability is a yet unknown parameter in these calculations. The assumption $w_{stick} \cong 1$ is probably far from reality. Laboratory experiments are necessary to get more reliable parameters for these interactions.

According to Hayashi et al. (1985) the corresponding sedimentation time for vertical settling can be derived from the above given equations, leading to

$$t_{sed} = t_K \frac{4}{\pi^{3/2}} \frac{1}{w_{stick} \varsigma (1 + \frac{8\rho_p}{w_{stick}\Sigma})r_0} \ln \frac{z_0 r}{z r_0}$$ (13)

where t_K is the Keplerian orbital period. Using the above given parameters for the planetary and the satellite systems, it can be derived that the typical sedimentation time is of the order of some 10^3 years for the planetary system, but it is only of the order of about 50 years for the satellite systems. The small sedimentation time in these systems is caused mainly by the assumed higher densities and the smaller scales in these systems. This time scale would be prolonged, if not a stable and more dense but a transient and thinner disk is assumed for these systems. The characteristic life-time of these transient pre-satellite disks is determined then by that of the preplanetary disk, feeding the pre-satellite disks. But, in every case, there forms a thin midplane-subdisk of dust. The above mentioned processes of self-gravitation may have become effective in the small scales in these dust subdisks, supporting the growth of larger bodies from the dust.

It is interesting to note in this context that growth processes are limited by the tidal action of the central body within the so-called "Roch-limit". Here, only small bodies can form and survive. Their interaction and erosion is probably an essential cause of the ring phenomenon, observed in the planetary satellite systems. The other source of single ring phenomena are probably decay processes of satellites. For more details, related to the origin of planetary rings the interested reader is referred to Harris (1984).

10. Collisional accretion towards larger bodies

Following the original ideas of Schmidt and Safronov (Safronov, 1972), the growth of larger bodies via the phase of km-sized planetesimals is assumed to be dominated by collisional accretion. The time scale of these processes can be estimated from the inverse of the collision frequency of colliding particles of cross section Q, number density N and average velocity $\bar{v}$

$$t_{acc} = \frac{1}{NQ\bar{v}}$$ (14)

With $H = \bar{v} / \Omega_K$, $\Sigma = 2H\rho$ and

$$N = \frac{\rho}{m} = \frac{\Sigma}{2mH} = \frac{\Sigma\Omega_K}{2m\bar{v}}$$ (15)

there follows for the characteristic time of accretional growth towards a planet of density ρ_{Pl} and Radius R_{Pl}

$$t_{acc} = \frac{4\rho_{Pl}R_{Pl}}{3\pi\Sigma}\, t_K \qquad (16)$$

To form a body of 1000km in radius in the inner planetary system, a time of about 10^7 years seems to be typical, but problems appear in the outer parts of the system, where this timescale increases over the age of the Solar system. Some processes of equipartitioning of energy between the larger and the smaller bodies have been discussed, and it has been shown by Wetherill (1991) that a runaway growth is able to shorten the longer time scales.

For the satellite systems, the above derived timescale is drastically shorter, it ranges between some hundred and thousand years (if the above given values for pre-satellite disks with relatively high density are used). So, collisional accretion can be assumed to be the basic growth mechanism for the larger bodies in all the four systems under consideration.

11. The References

Cameron, A.G.W.(1978) Physics of the primitive Solar accretion disk, *Moon and Planets*, 98,5-40

Harris, A.W.(1984) Origin of Planetary Rings, in *Planetary Rings*, (R. Greenberg and A. Brahic ,Eds., 641-659, U. of Arizona Press, Tucson

Hayashi, C., Nakazawa, K., and Nakagawa Y. (1985) Formation of the Solar System, in *Protostars and Planets II*, eds. Black, D.C. and M.S. Matthews, 1100-1153, University of Arizona Press.

Lang, K.R.(1992) *Astrophysical Data*, Planets and Stars, Springer Verlag

Lin, D.N.C. and J. Papaloizou (1985) On ther dynamical origin of the Solar system, in *Protostars and Planets II*, eds. D.C. Black and M.S. Matthews, 981-1072, University of Arizona Press

Lissauer, J.J. and J.N. Cuzzi (1985) Rings and Moons:Clues to understanding the Solar nebula, in *Protostars and Planets II*, eds. D.C. Black and M.S. Matthews, 920-958, University of Arizona Press

Safronov, V.S. (1972) in *Evolution of the Protoplanetary Cloud and Formation of the Earth and Planets*, Moscow, Nauka, 1969), NASA-TT-F-667,1972

Weidenschilling, S.J. and J.N. Cuzzi (1993) Formation of Planetesimals in the Solar Nebula, in *Protostars and Planets III*, eds. Levy, E.H. and J.I. Lunine, pp.1031-1060, University of Arizona Press

Wetherill, G.W. (1991) Formation of the Terrestrial Planets from Planetesimals, in *Planetary Sciences, American and Soviet Research*, ed. T. Donahue, Washington, Ntl. Acad. of Sciences Press

FROZEN FIELDS

L. L. HOOD
Lunar and Planetary Laboratory
University of Arizona, Tucson, Arizona 85721

Abstract. Magnetic fields due to permanent magnetization of planetary crusts and interiors have been clearly detected only for the Earth and Moon. However, they are likely to be a ubiquitous property of silicate and partially silicate objects in the solar system. An indication that this is true is the recent indirect evidence from the Galileo flybys that the asteroids Gaspra and Ida have intrinsic magnetic fields. Lunar paleomagnetism differs substantially from terrestrial paleomagnetism in part because the dominant ferromagnetic carriers are metallic Fe-Ni grains rather than iron oxides such as magnetite. The distribution of metallic iron remanence carriers on the Moon is influenced strongly by impact processes. In addition, large-scale lunar impacts may have produced transient magnetic fields capable of imparting magnetization with or without a former core dynamo. An unresolved issue of lunar paleomagnetism is the origin of swirl-like albedo markings associated with the strongest magnetic anomalies detected from orbit. The interpretation of solar wind magnetic field perturbations during the Gaspra and Ida flybys as due to intrinsic asteroidal magnetic fields has been supported by detailed magnetohydrodynamic simulations. The inferred magnetization limits for Gaspra are consistent with a wide variety of meteorite types and do not allow firm constraints to be imposed on Gaspra's bulk composition.

1. Introduction

The term "frozen fields" refers to magnetic fields of solar system bodies that are due to permanent magnetization of surface materials or of the interior. Among solar system objects, only the Earth and the Moon have been surveyed magnetically in sufficient detail to allow the direct detection of frozen fields. However, because of the ubiquitous presence of magnetization in meteorites and the widespread existence of internally generated planetary magnetic fields, it is likely that frozen fields are a common property of most silicate solar system bodies. The recent Galileo flybys of the asteroids Gaspra and Ida have yielded indirect evidence for frozen fields in the form of perturbations of the interplanetary magnetic field caused by the interaction of a magnetized plasma (the solar wind) with a magnetized body. By analogy with the lunar case, it is very likely that the planet Mercury possesses crustal magnetization. However, as will be seen below, it is unlikely that this planet's weak intrinsic dipolar magnetic field is due to frozen fields. Although the Venusian crust is probably too hot to allow the retention of significant permanent magnetization, the planet Mars may be expected to have strong crustal magnetic anomalies if a core dynamo magnetic field existed in the past. (For recent studies of the possible existence of a weak Martian magnetic field, including possible evidence for a magnetic anomaly in the Tharsis region, see Möhlmann [1992] and Möhlmann et al. [1991].) The Martian satellites Phobos and Deimos are likely to be weakly magnetized based on comparisons with the Moon, Gaspra, and Ida. Finally, it may be expected that outer planet satellites having at least a partially silicate composition will exhibit frozen fields due either to exposure to external

Earth, Moon, and Planets 67: 131–142, 1995.

planetary magnetic fields or to local surface processes that may have generated transient magnetic fields (e.g., impacts).

The character of crustal or internal magnetization expected for a given silicate body in the solar system will obviously depend heavily on the body's history and on the nature of the dominant magnetic remanence carriers. For dry silicate objects such as the Moon, Mercury, and most asteroids, metallic iron will be the dominant ferromagnetic carrier because of the strongly reducing conditions. For small undifferentiated (previously unmelted) asteroids, a magnetization of the interior similar to that observed for meteorites (possibly acquired primordially in the solar nebula) can be hypothesized. For larger differentiated objects such as the Moon and Mercury, any internal magnetization must be post-primordial and will be limited to an outer shell where temperatures are presently low enough to allow magnetization to be retained. Because metallic iron is produced by reduction of pre-existing iron silicates during meteoroid impacts, impact processes will play a major role in controlling the distribution of magnetized materials on such objects. Furthermore, as will be seen, impacts may have produced transient magnetic fields that could have been the source of at least part of any observed crustal magnetization. Finally, objects such as the Earth and Mars with atmospheres and oxidizing conditions will have iron oxides (e.g., magnetite) as the dominant remanence carriers. Magnetization of crustal materials will be due primarily to thermal remanence acquired by slow cooling of geologic units such as magmatic intrusions in the presence of a steady global magnetic field generated in the planetary core (see, e.g., Stacey and Banerjee [1974] and Hahn [1971]). In this paper, a brief survey will be presented of non-terrestrial frozen fields detected either directly or indirectly so far in the course of planetary exploration. As will be seen, the lunar data are by far the most complete but yield at least as many new questions as answers. Among these questions are: (1) Are large-scale impacts on planetary surfaces capable of generating transient magnetic fields that would produce magnetization even in the absence of a core dynamo magnetic field? and (2) What is the origin of swirl-like albedo markings that are associated with many of the strongest lunar magnetic anomalies? Although it has been suggested that Mercury's weak intrinsic magnetic field could be due to remanent magnetization, the prevailing evidence indicates otherwise, as will be discussed. Indirect evidence exists that the asteroids Gaspra and Ida probably have frozen fields consistent with the observed magnetization intensities for many meteorite classes. However, it will be shown that the derived limits on the magnetization intensity of Gaspra are not sufficient to significantly constrain the composition (metallic or silicate) of this asteroid.

2. The Moon

Although flyby and orbiter measurements during the 1960's demonstrated that the Moon had no large-scale intrinsic magnetic field, more detailed measurements obtained during the Apollo manned landing missions revealed a pervasive and variable magnetization of the lunar crust (for reviews, see Fuller [1974]; and Fuller and Cisowski [1987]). In addition to surface magnetic field measurements [Dyal et al., 1974], orbital measurements obtained at low altitudes (∼20 to 100 km) with the Apollo 15 and 16 Particles and Fields subsatellites identified crustal magnetic anomalies with scale sizes of hundreds of kilometers [Coleman et al., 1972; Russell et al., 1975; Lin, 1979; Hood et al., 1981]. Both direct orbital magnetometer measurements and indirect electron reflection measurements were used to map the distribution and intensity of lunar surface fields at latitudes less than about 30°.

The top panel of Figure 1 (from Lin et al. [1988]) shows the distribution of surface magnetic fields as mapped using the electron reflection technique. (Note that this map is centered on the central far side.) It should be emphasized that the stated field amplitudes (> 1.6 nT and > 3.2 nT) are qualitative estimates only. Actual surface fields measured at the Apollo landing sites range up to 327 nT and extrapolations of orbital magnetometer data suggest that surface fields exceeding 1000 nT (0.01 Gauss) may exist within limited regions. Although this map has a relatively low resolution, it does provide a rough description of the global distribution of anomalies. Specifically, the largest anomalies tend to be on the far side; especially large concentrations occur near 20°N, 90°E and near 30°S, 160°E. The dashed circles on the map indicate the locations of the antipodes of major lunar ringed impact basins. It is seen that the two largest concentrations of anomalies occur near the antipodes of the two youngest large impact basins: Orientale and Imbrium. Two additional concentrations of anomalies occur near the antipodes of the two next youngest large basins: Serenitatis and Crisium. (For further documentation including evidence from alternate field mapping methods and statistical analysis, see Lin et al. [1988]). A model for the origin of magnetic anomalies antipodal to lunar impact basins has been partially developed theoretically [Hood and Huang, 1991] and will be discussed further below.

In addition to the observed correlation of large magnetic anomaly concentrations with impact basin antipodes, a number of other observational constraints suggest that impact processes have played an important role in producing the observed magnetization. From orbital data, low-altitude passes across the lunar near side by the Apollo 16 subsatellite magnetometer showed that fields are relatively weak across the maria but are larger over exposures of the Fra Mauro Formation (primary Imbrium basin ejecta) and over the Cayley Formation (secondary basin ejecta) [Hood et al., 1979]. Surface magnetometer measurements also showed relatively weak fields at Apollo mare sites but stronger fields at highland sites including the Apollo 16 site, dominated by the Cayley Formation, where the strongest surface fields were measured [Strangway et al., 1973a]. Sample studies had previously shown that mare basalts contain relatively little of the microscopic metallic iron remanence carriers while impact-produced breccias and soils contain much more metallic iron [Strangway et al., 1973b]. Thus, it is not surprising that the maria are magnetically relatively weak compared to the highlands and compared to basin ejecta materials such as the Fra Mauro Formation.

It is still unclear whether a former lunar core dynamo or transient fields generated during impacts (or both) are responsible for providing lunar magnetizing fields. Based on mean density, moment-of-inertia, and bulk composition

134

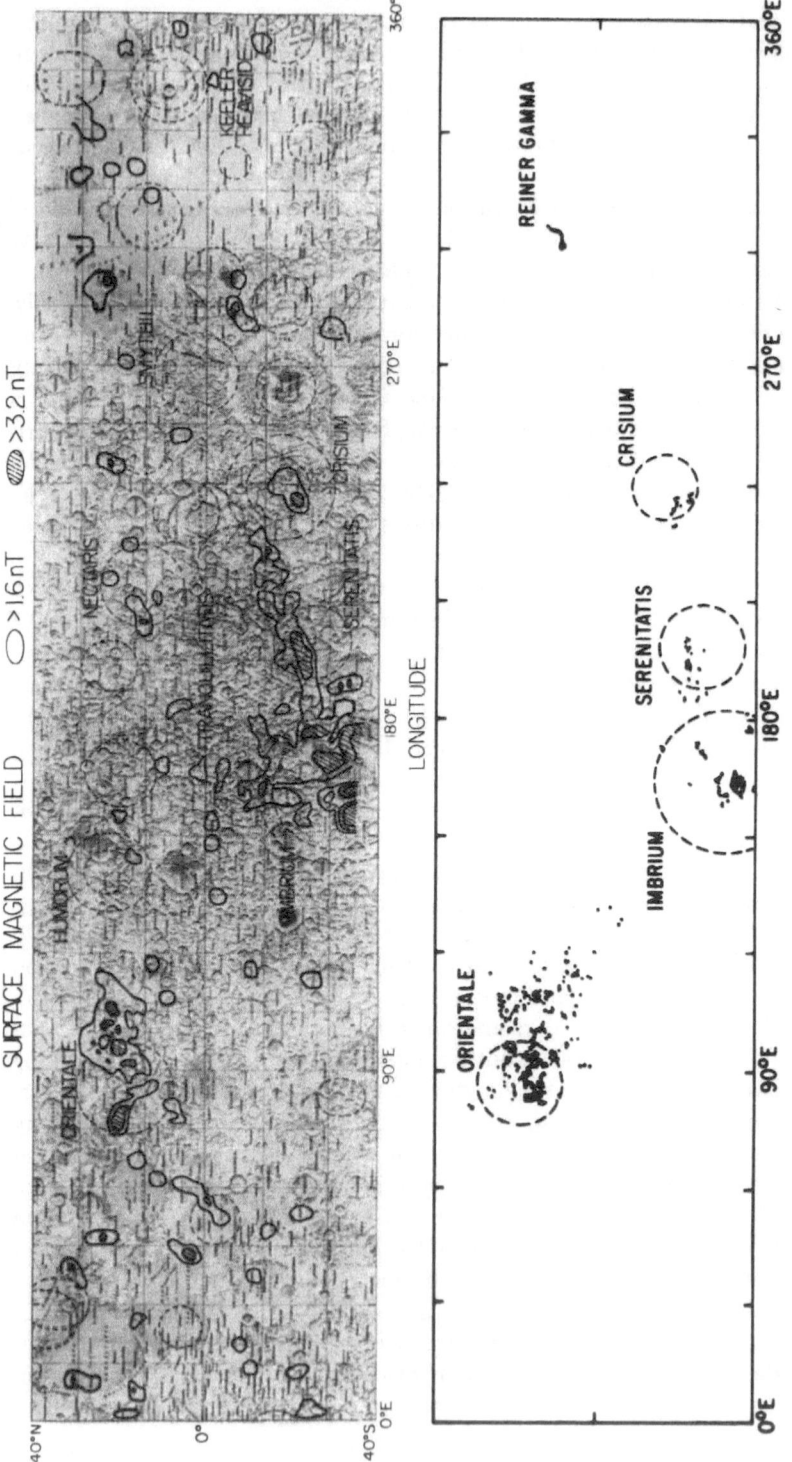

Figure 1. Top panel: Distribution of lunar surface magnetic fields estimated by the electron reflection technique based on Apollo 15 and 16 subsatellite energetic charged particle data. (from Lin et al. [1988]). The dashed circles are centered on the antipodes of ringed impact basins. Bottom panel: Distribution of swirl-like albedo markings of the Reiner Gamma class based in part on original maps by Schultz and Srnka [1980].

constraints, it is very likely that a small metallic core with a radius near 400 km exists in the Moon [Hood and Jones, 1987]. However, it is unknown whether such a core generated a dynamo magnetic field early in the Moon's history. Compilations of paleointensity estimates for returned lunar samples suggest relatively large (~ 1 Gauss) magnetizing fields between about 3.6 and 3.8 aeons [Cisowski et al., 1983; Cisowski and Fuller, 1986]. This might be suggestive of a core dynamo. However, the magnetic properties of reduced Fe-Ni remanence carriers in lunar samples (as well as most meteorites) remain only partially understood [Wasilewski, 1988]. In addition, significant paleointensities were also estimated for samples with ages of less than 1 aeon. In particular, an Apollo 17 impact glass sample with an age of

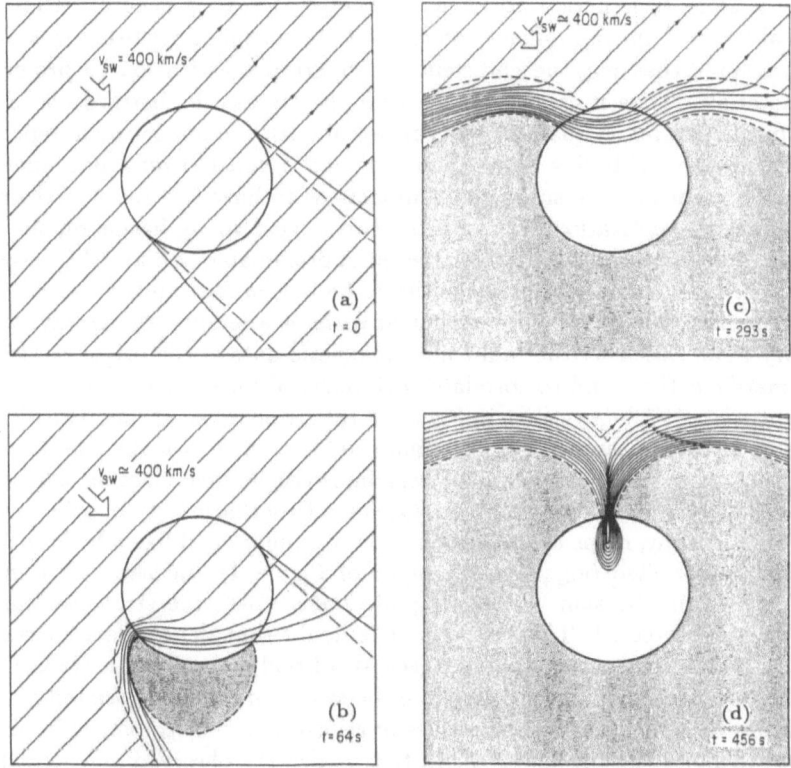

Figure 2. Model of the magnetohydrodynamic interaction of an expanding basin-scale impact plasma cloud with a simplified solar wind consisting of a constant-density plasma and magnetic field assumed to be oriented perpendicular to the Sun-Moon line (from Hood and Huang [1991]).

< 200 Myr yielded a paleointensity estimate of 2500 nT [Sugiura et al., 1979]. The latter field estimates for the relatively recent lunar past suggest that transient fields may have been generated by, e.g., the impact that produced the glass sample.

The major characteristic of meteoroid impacts at velocities greater than about 10 km/s that admits the possibility of transient field generation is the production of a cloud of partially ionized, hot silicate vapor in addition to solid and molten ejecta. This impact plasma cloud expands rapidly with a speed comparable to the impact velocity [Melosh, 1989] and interacts strongly with surrounding ambient magnetic fields [Hide, 1972; Srnka, 1977; Hood and Vickery, 1984]. Transient magnetic fields generated in laboratory-scale impacts have recently been documented [Crawford and Schultz, 1988; 1991]. In the case of a large basin-forming impact, models suggest that a compression of the ambient magnetic field will occur antipodal to the impact point as illustrated in Figure 2 (from Hood and Huang [1991]). During the same time period as when the field compression occurs, seismic compressional waves from the impact converge at the antipode resulting in transient shock pressures that may be sufficient to impart magnetization to pre-existing lunar materials. Thus it can be argued that the large-scale distribution of lunar crustal magnetization is qualitatively consistent with the impact field hypothesis. Antipodal regions of basins older than Serenitatis and Crisium may have been sufficiently "gardened" by subsequent impacts as to have lost most of their original coherent magnetization. Because the two largest young basins on the Moon (Orientale and Imbrium) have ages between approximately 3.6 and 4.0 aeons, the tendency for lunar sample paleointensities to be largest for samples formed during this time period may possibly be explained without a temporary core dynamo. A secondary issue raised by the lunar paleomagnetic data is the origin of swirl-like albedo markings that tend to correlate with many of the strongest orbital anomalies [Hood et al., 1979]. As shown in Figure 3, the strongest single anomaly detected during low-altitude passes across the equatorial near side with the Apollo 16 subsatellite magnetometer was found to correlate with Reiner Gamma, an unusual curvilinear albedo marking on western Oceanus Procellarum. Using the available orbital photography, maps of the distribution of similar swirl-like albedo markings have been constructed (e.g., bottom panel of Figure 1) and show that most of the swirls exist in the same basin antipode zones where the strongest magnetic anomalies were detected. Thus there is little doubt that the swirls are associated with strong lunar magnetic anomalies (see also Hood and Williams [1989]).

One proposed model for the origin of the swirls and their associated magnetic anomalies is that they represent residues of relatively recent cometary impacts on the Moon [Schultz and Srnka, 1980]. In this model, the observed magnetization is a consequence of compression of the interplanetary magnetic field by ionized gas from the cometary coma in the presence of shock effects of the impact. However, cometary impacts would be expected to occur nearly at random on the Moon rather than occurring preferentially antipodal to young large basins. Therefore, unless the correlation with young basin antipodes is coincidental (unlikely according to Lin et al. [1988]), the cometary impact hypothesis seems disfavored.

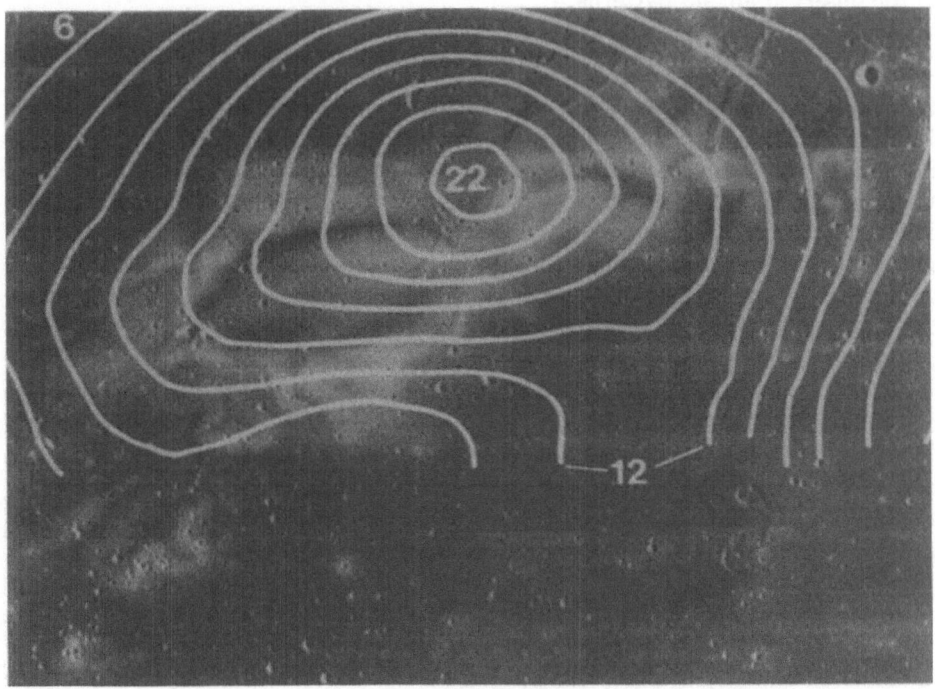

Figure 3. Superposition of a contour map of the lunar crustal magnetic field (nT) as derived from Apollo 16 subsatellite magnetometer data onto a photograph of Reiner Gamma, an unusual curvilinear albedo marking on western Oceanus Procellarum. The albedo marking is approximately 60 km in length.

A second model for the origin of the swirls proposes that they are a consequence of deflection of the solar wind ion bombardment by the magnetic anomaly fields [Hood and Schubert, 1980; Hood and Williams, 1989]. Briefly, this model assumes that solar wind hydrogen is at least partially involved in the process that darkens lunar surface materials with time (typified by the darkening of crater rays after periods of the order of 1 aeon). The primary darkening process is known to be micrometeoroid impacts that produce microscopic metallic iron in glassy aggregate particles called agglutinates. However, the rate of metal generation in agglutinates may be a partial function of the amount of reducing solar wind hydrogen that is implanted in the uppermost regolith [Housley et al., 1973]. Thus, areas of the surface that are relatively shielded from the solar wind ion bombardment by locally strong crustal fields may have their original higher albedos preserved compared to the rest of the lunar surface. Model calculations involving numerical simulations of solar wind ion deflection by crustal magnetic fields have shown that the curvilinear shapes of the observed albedo markings can in principle be explained by the magnetic ion deflection model [Hood and Williams, 1988; Hood, 1992]. It is also possible that the settling of fine secondary crater ejecta containing paramagnetic iron may have been modified by crustal magnetic fields contributing to the observed albedo patterns.

3. Mercury

Based on data from the two Mariner 10 flybys, it is well established that the planet Mercury possesses a weak dipolar magnetic field with a moment of $\sim 300^{+200}_{-100}$ nT-R_M^3 (1 R_M = 2439 km) inclined by an angle of roughly $10°$ to the planetary spin axis [Connerney and Ness, 1988]. Although an internally generated hydromagnetic dynamo in the planetary core is almost universally accepted as the most probable source of this field, one of the alternate possibilities that has been mentioned in the literature is permanent magnetization of the planetary crust. By analogy with the lunar case, it would not be surprising at all if crustal magnetization is indeed present at Mercury. However, as reviewed by Schubert et al. [1988], there are many problems with attempting to explain the observed global dipole moment as due to crustal magnetization alone.

First of all, typical Mercury surface fields of $\sim$ 300 nT are much larger than that of the crustal field component on Earth ($\sim$ 10 nT) and that typically found on the Moon ($\sim$ 50 nT). Moreover, in both the terrestrial and lunar cases, magnetization directions are sufficiently random that there is no global crustal magnetic dipole moment. Thus, Mercury would be anomalous if its crustal magnetization produced a substantial dipole moment. Second, the time scale for growth of a cold magnetizable lithospheric shell on Mercury ($\sim$ 100 Myr) is much longer than the time scale for the steady existence (without reversals) of any internally or externally generated magnetic field. For example, the terrestrial dipole moment reverses polarity at typical time intervals of only a few hundred thousand years. If the same time scale is approximately appropriate for reversals of a former core dynamo on Mercury, then the field would have reversed about 500 times during the formation of a magnetizable shell. Considering likely lateral variations in susceptibility and possible polar wander together with field reversals, it is doubtful that any coherent magnetization leading to a global dipole moment would remain.

4. Gaspra and Ida

Recently, indirect evidence that asteroids may possess significant frozen fields resulting from internal magnetization has been obtained as a result of the Galileo spacecraft flybys of Gaspra and Ida in 29 October 1991 and 28 August 1993, respectively [Kivelson et al., 1993a,b]. Both 951 Gaspra and 243 Ida have spectral properties that put them in the S taxonomic class. Gaspra has a mean radius of about 7 km while Ida is at least 10 times more massive with a mean radius of about 30 km. Both objects are elongated and irregularly shaped. During both the Gaspra flyby (closest approach about 1600 km) and the Ida flyby (closest approach about 2400 km), the Galileo magnetometer observed substantial rotations of the interplanetary magnetic field that were probably caused by the interaction of the solar wind with a magnetized object. Because the measurements and interpretation of the Ida results are currently relatively incomplete, the discussion below is limited to the Gaspra results.

As described by Kivelson et al. [1993a], the interplanetary magnetic field apparently rotated toward Gaspra beginning one minute before closest approach and rotated back toward its original orientation two minutes after closest approach. According to the analysis of Kivelson et al., this magnetic field perturbation can be interpreted as a "ripple" in the interplanetary field caused by draping of the solar wind magnetic field around a small magnetospheric obstacle. The probable validity of this interpretation is supported by a detailed magnetohydrodynamic simulation of the interaction by Baumgärtel et al. [1994]. The latter simulation approximately reproduces the Galileo magnetic signature if Gaspra has a magnetic dipole moment of the order of 10^{14} Amp-m^2 tilted by an angle of about 45° to the solar wind flow direction. Kivelson et al. estimated that Gaspra's magnetic moment is between 6×10^{12} and 2×10^{14} Amp-m^2. (Note: 1 Amp-m^2 = 10^3 Gauss-cm^3) Dividing by the approximate volume of Gaspra (about $1.33*\pi*(7$ km$)^3$) yields a mean magnetization intensity of between about 4 and 140 Amp-m^{-1}. Assuming a nominal density of 4000 kg m^{-3} then implies limits on the magnetization per unit mass of 0.001 and 0.035 Amp-m^2 kg^{-1}.

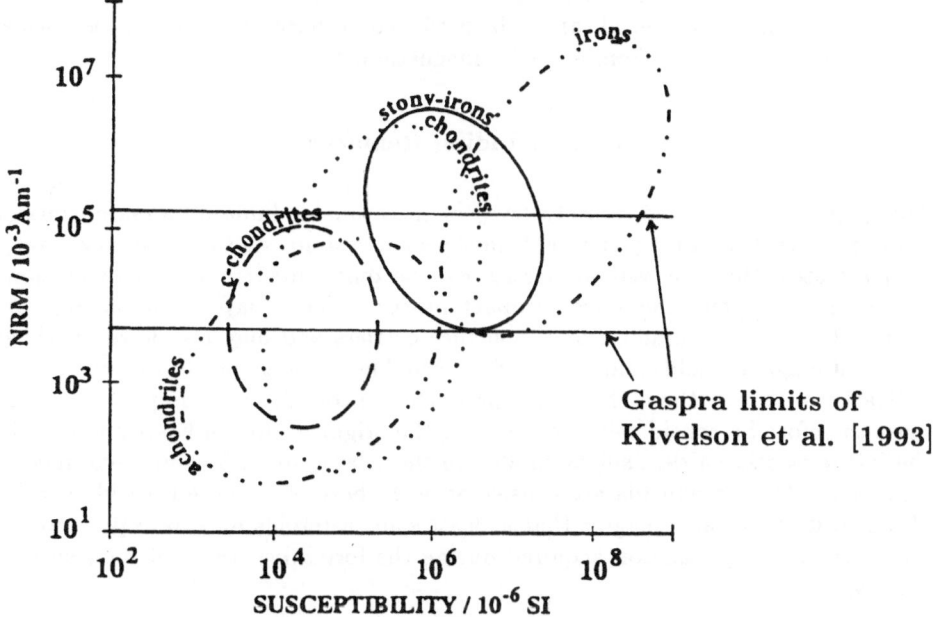

Figure 4. Comparison of the Gaspra magnetization limits of Kivelson et al. [1993] with Finnish Antarctic meteorite magnetization data (plotted against susceptibility) of Pesonen et al. [1993]. The ovals represent approximate bounds of the data distribution for the indicated meteorite types.

A major unresolved issue relating to S-type asteroids is whether they are parent bodies of ordinary chondrite meteorites or of the stony-iron meteorites. In the former case, they are undifferentiated and have relatively little metal while in the latter case they have been differentiated into metal and silicate components during an early heating event. Therefore, it is of interest to investigate whether the inferred Gaspra magnetic moment limits impose any significant constraints on Gaspra's composition. This is especially true in view of the fact that the Galileo flyby was not close enough to allow an estimate of Gaspra's mass (and hence density). Using a limited sample of meteorite magnetization measurements, Kivelson et al. [1993b] concluded that the inferred magnetization limits are "in the range observed for iron meteorites and highly magnetized chondrites". Such a conclusion would suggest a significant metallic component for Gaspra supporting the stony-iron model of S-type asteroids (e.g., Kerr [1993]). However, as shown in Figure 4, compositional inferences based on the Galileo magnetic field signature may be questionable. The labeled ovals summarize magnetization measurements vs. susceptibility for a large sample of 368 Antarctic meteorites compiled by Pesonen et al. [1993]. The horizontal lines are the upper and lower limits for Gaspra inferred from the Galileo magnetic field signature. As can be seen, these limits are formally consistent with virtually all meteorite types ranging from irons to stony irons to ordinary chondrites, carbonaceous chondrites, and achondrites. Both H and L ordinary chondrites can be accommodated. A similar conclusion can be drawn on the basis of Russian meteorite data summarized by Sonett [1978] although the Finnish data of Pesonen et al. are to be preferred because the collected meteorites were carefully protected from artificial magnetic fields.

5. Concluding Remarks

This paper has presented a brief survey of the available limited measurements of non-terrestrial frozen magnetic fields in the solar system. As the lunar case amply demonstrates, the observed properties can be quite different from expectations based on terrestrial experience. Impact processes have played a dominant role in distributing the metallic iron remanence carriers and may also have provided transient magnetic fields and shock effects capable of imparting the observed magnetization with or without the former presence of a core dynamo. Impact processes must therefore be considered in evaluating the origins of frozen fields observed in the future at other airless silicate bodies in the solar system. The inferred magnetizations of Gaspra and Ida are consistent with those observed for a wide variety of meteorite types and suggest that at least some asteroids may have preserved a "primordial" magnetization acquired during the formative stages of solar system history.

Acknowledgment
I thank Dr. Peter Wasilewski of Goddard Space Flight Center for referring me to the Finnish meteorite magnetization data used in Figure 4.

References

Baumgärtel, K., Sauer K., and Bogdanov, A. (1994) A magnetohydrodynamic model of solar wind interaction with asteroid Gaspra, *Science* **263**, 653-655.

Cisowski, S. M., Collinson, D. W., Runcorn, S. K., Stephenson, A., and Fuller, M. (1983) A review of lunar paleointensity data and implications for the origin of lunar paleomagnetism, *Proc. Lunar Planet. Sci. Conf. 13th, J. Geophys. Res.* **88**, Suppl. A691-A704.

Cisowski, S.M. and Fuller M. (1986) Lunar paleointensities via the IRMs normalization method and the early magnetic history of the moon, in W.K. Hartmann, R.J. Phillips, and G.J. Taylor (eds.), *Origin of the Moon*, Lunar and Planetary Institute, Houston, pp. 411-424.

Coleman, P. J., Jr., Schubert, G., Russell, C. T., and Sharp, L. R. (1972) Satellite measurements of the Moon's magnetic field: A preliminary report, *The Moon* **4**, 419-429.

Connerney, J. E. P. and Ness, N. F. (1988) Mercury's magnetic field and interior, in F. Vilas, C. R. Chapman, and M. S. Matthews (eds.), *Mercury*, pp. 494-513.

Crawford, D. and Schultz, P. (1988) Laboratory observations of impact-generated magnetic fields, *Nature* **336**, pp. 50-52.

Crawford, D. and Schultz, P. (1991) Laboratory investigations of impact-generated plasma, *J. Geophys. Res.* **96**, 18807-18819.

Dyal, P., Parkin, C. W., and Daily, W. D. (1974) Magnetism and the interior of the Moon, *Rev. Geophys. Space Phys.* **12**, 568-591.

Fuller, M. (1974) Lunar magnetism, *Rev. Geophys. Space Phys.* **12**, 23-70.

Fuller, M. and Cisowski, S. (1987) Lunar Paleomagnetism, in J. A. Jacobs (ed.), *Geomagnetism*, Academic Press, London, pp. 307-456.

Hahn, A. G. (1971) Types of magnetic anomalies measured on land and general aspects of their geological meaning, in A. J. Zmuda (ed.), *World Magnetic Survey 1957-1969*, IUGG, Paris, pp. 134-142.

Hide, R. (1972) Comments on the moon's magnetism, *The Moon* **4**, 39.

Hood, L. L. (1992) Lunar magnetic fields: Implications for utilization and resource extraction, *J. Geophys. Res.* **97**, 18275-18284.

Hood, L.L., Coleman, P.J. Jr., and Wilhelms, D.E. (1979) The Moon: Sources of the crustal magnetic anomalies, *Science* **204**, 53-57.

Hood, L. L. and Huang, Z. (1991) Formation of magnetic anomalies antipodal to lunar impact basins: Two-dimensional model calculations, *J. Geophys. Res.* **96**, 9837-9846.

Hood, L. L. and Jones, J. (1987) Geophysical constraints on lunar bulk composition and structure: A reassessment, *J. Geophys. Res.* **92**, supplement, E396-E410.

Hood, L. L., Russell, C. T., and Coleman, P. J., Jr. (1981) Contour maps of lunar remanent magnetic fields, *J. Geophys. Res.* **86**, 1055-1069.

Hood, L.L. and Schubert G. (1980) Lunar magnetic anomalies and surface optical properties, *Science* **208**, 49-51.

Hood, L.L. and Vickery A. (1984) Magnetic field amplification and generation in hypervelocity meteoroid impacts with application to lunar paleomagnetism, *Proc. Lunar Planet. Sci. Conf. 15th*, in *J. Geophys. Res.* **89**, C211-C223.

Hood, L. L. and Williams, C. R. (1989) The lunar swirls: Distribution and possible origins, *Proc. Lunar Planet. Sci. Conf. 19th*, Lunar and Planetary Institute, Houston, pp. 99-113.

Housley, R. M., Grant, R. W., and Paton, N. E. (1973) Origin and characteristics of excess Fe metal in lunar glass welded aggregates, *Proc. Lunar Sci. Conf. 4th*, pp. 2737-2749.

Kerr, R., (1993) Magnetic ripple hints Gaspra is metallic, *Science* **259**, 176.

Kivelson, M., Bargatze, L., Khurana, K., Southwood, D., Walker, R., and Coleman, P., Jr. (1993a) Magnetic signatures near Galileo's closest approach to Gaspra, *Science* **261**, 331-334.

Kivelson, M., Khurana, K., Russell, C., Southwood, D., Walker, R., and Wang, Z. (1993b) Ida Flyby: First Results from the Galileo Magnetometer, *EOS*, Fall Meeting Abstracts, p. 384.

Lin, R. P., Anderson, K. A., and Hood, L. L. (1988) Lunar surface magnetic field concentrations antipodal to young large impact basins, *Icarus* **74**, 529-541.

Lin, R. P., (1979) Constraints on the origins of lunar magnetism from electron reflection measurements of surface magnetic fields, *Phys. Earth Planet. Inter.* **20**, 271-280.

Melosh, H. J., (1989) *Impact Cratering*, Oxford University Press, New York.

Möhlmann, D. (1992) The question of a Martian planetary magnetic field, *Adv. Space Res.*

12, 213-217. Möhlmann, D., Riedler, W., Rustenbach, J., Schwingenschuh, K., Kurths, J., Motschmann, U., Roatsch, T., Sauer, K., and Lichtenegger, H. (1991) The question of an internal Martian magnetic field, *Planet. Space. Sci.* **39**, 83-88.

Pesonen, L. J., Terho, M., and Kukkonen, I. (1993) Physical properties of 368 meteorites: Implications for meteorite magnetism and planetary geophysics, *Proc. NIPR Symp. Antarct. Meteorites* **6**, 401-416.

Russell, C. T., Coleman, P. J., Jr., Fleming, B. K., Hilburn, L., Ioannidis, G., Lichtenstein, B. R., and Schubert, G. (1975) The fine scale lunar magnetic field, *Proc. Lunar Sci. Conf. 6th*, 2955-2969.

Schultz, P.H. and Srnka, L.J. (1980) Cometary collisions on the moon and Mercury, *Nature* **284**, 22–26.

Schubert, G., Ross, M., Stevenson, D., and Spohn, T. (1988) Mercury's thermal history and the generation of its magnetic field, in F. Villas, C. Chapman, and M. S. Matthews (eds.), *Mercury*, University of Arizona Press, Tucson, pp. 429-460.

Sonett, C. P. (1978) Evidence for a primordial magnetic field in the meteorite parent body era, *Geophys. Res. Lett.* **5**, 151-154.

Srnka, L.J. (1977) Spontaneous magnetic field generation in hypervelocity impacts, *Proc. Lunar Sci. Conf. 8th*, Lunar and Planetary Institute, Houston, pp. 792-795.

Stacey, F. D., and Banerjee, S. K. (1974) *The Physical Principles of Rock Magnetism*, Elsevier, New York.

Strangway, D. W., Gose, W., Pearce, G., and McConnell, R. K. (1973a) Lunar magnetic anomalies and the Cayley formation, *Nature* **246**, 112-114.

Strangway, D. W., Sharpe, H., Gose, W., and Pearce, G. (1973b) Magnetism and the history of the Moon, in C. D. Graham, Jr. and J. J. Rhyne (eds.), *Magnetism and Magnetic Material - 1972*, American Institute of Physics, New York, pp. 1178-1187. Sugiura, N., Wu, Y. M., Strangway, D. W., Pearce, G. W., and Taylor, L. A. (1979) A new magnetic paleointensity value for a young lunar glass, *Proc. Lunar Planet. Sci. Conf. 10th*, 2189-2198.

Wasilewski, P. J. (1988) Magnetic characterization of the new magnetic mineral tetrataenite and its contrast with isochemical taenite, *Phys. Earth Planet. Inter.* **52**, 150-158.

PLANETARY DYNAMOS

E. H. Levy
Department of Planetary Sciences
Lunar and Planetary Laboratory
The University of Arizona
Tucson, Arizona 85721

Abstract

Planetary magnetic fields are thought to be generated by magnetohydrodynamic dynamos acting in the convecting, electrically conducting fluid cores of these cosmic bodies. Similar processes are believed to produce a wide variety of other cosmic magnetic fields, including the fields of the Sun and stars. At present, we understand the basic physical processes involved in dynamo magnetic field generation. However, a detailed understanding still eludes us both because of continuing uncertainty about planetary interior structures, properties, and fluid motions, and because of our still primitive capacity for dealing with the complex, nonlinear dynamical processes involved in the fully elaborated dynamo process.

1. Introduction

Despite theoretical arguments suggesting the existence of free magnetic charge in our universe, such particles have not yet been discovered. So far as we know, all magnetic fields thus far observed are produced by the motion of electric charge—known as electric current. The large scale magnetic fields associated with macroscopic objects involve the correlated macroscopic motion of many electric charges to make large scale electric currents. In the case of planets the main magnetic-field-producing electric currents are confined

Earth, Moon, and Planets **67**: 143–160, 1995.
© 1995 *Kluwer Academic Publishers.*

within the planets. In addition to these interior main-field-producing planetary currents, weaker electric currents are induced by the interaction between planetary magnetic fields and the solar wind that blows by the planet. These latter currents—for the most part exterior to the body of the planet—distort the field strongly at large planetocentric distances. Here we are concerned entirely with the internal electric currents that produce the main planetary magnetic fields.

The correlated electrons and ions moving to produce the large scale magnetic-field-generating electric currents are subject to the disruptive influence of collisions with the matter through which they move. This serves to disrupt the coherent motion, stopping the electric current and dissipating the magnetic field, converting the energy into heat—a phenomenon that we describe as electrical resistance. Thus the persistence of a planetary magnetic field for times long in comparison to this Joule dissipation time requires the action of some mechanism for regenerating the the electric current. This is in contrast to the phenomenon of so-called *permanent magnetism*, in which there is no motion of electric charge over macroscopic distances. In permanent magnetism, although electron motions are correlated over macroscopic distances, the motions themselves are confined within individual atoms. Thus the collisional dissipation that affects macroscopic currents does not come into play. Left to itself, and at low enough temperatures, permanent magnetism may persist indefinitely.

A magnetohydrodynamic (MHD) dynamo is a physical process through which motion of an electrically conducting fluid in the presence of a magnetic field produces electric currents that reinforce the original magnetic field, and counteracts the dissipation or decay that results from electrical resistivity or related dissipative phenomena.

1.1 THE PLANETARY MAGNETIC FIELDS

Eight planets have been scrutinized for the existence of magnetic fields. Six of those are known to have substantial fields: Mercury, Earth, Jupiter, Saturn, Uranus, and Neptune. The remaining two,

Venus and Mars, have weak internal magnetic fields, if they have any at all at the present time. Earth's Moon shows a patchwork of surface magnetization, at least some of which likely preserves a record of an earlier internal lunar magnetic field. There is indirect evidence of magnetization in asteroids, and primitive meteorites also contain a record of magnetization from the early years of the Solar System. Asteroid and meteorite magnetizations may record either a large-scale magnetic field present during the time that the material accumulated, or it may record magnetic fields generated within the parent bodies themselves; at this point, there is little to choose among these two possible origins.

The surface and external magnetic fields of planets have predominantly dipolar characters. To some extent, this is a result of the fact that the dipole is the lowest-order component of a vacuum field, and it falls as r^{-3}; all of the other multipole components fall more rapidly with distance from the planet's center. Thus it is generally true that the "surface" of the magnetic-field-generating region within the planet, the magnetic field deviates far more strongly from dipolar structure than does the external field—having a substantially larger admixture of higher order components. Indeed, the properties of the magnetic field at the surface of the field-generating "core" are more pertinent to understanding the field's generation than are the properties observed at or above the planet's surface, which depend on the somewhat incidental distance between the surface the core. Given adequate data and information about the planet, a reasonable job could be done of inverting from the externally observed field to the core-surface field. However, uncertainties about interior structures and dynamical behavior limits the confidence that one can place in such inversions.

Of course analysis of adequate external data can give the multipole expansion of the field, which is itself independent of the uncertainties in planet properties—at least outside of the highly electrically conducting regions of the planet. However, it is important to emphasize the importance of knowing the actual field strengths within the generation region. It is the actual field strength that determines the Lorentz force, which enters the fluid dynamical equations, so that the field strength within the generating region is a much more fundamental quantity than the magnetic moment(s).

Despite the uncertainties, a broad picture can be drawn of the known planetary magnetic fields. Leaving aside Mercury, all highly magnetized planets—Earth, Jupiter, Saturn, Uranus and Neptune— seem to have magnetic field intensities in their field-generating regions of the order of 1–10 Gauss. Mercury is the exception, with a field of the order of 10^{-2} Gauss. It is interesting to note that, of the magnetized highly magnetized planets, all but Mercury are rapid rotators, with rotation periods of the order of a day. Mercury rotates only once in 57 days. This is probably a significant fact (even admitting uncertainty about whether Mercury's magnetic field originates in a process similar to that which exists in the other planets). Inasmuch as it is the Coriolis force that organizes the fluid motion, and drives those correlated components of the motion that facilitate effective field generation, and it is the Lorentz force that ultimately must limit the generation and strength of the resulting magnetic field, it is not surprising to see a correlation between magnetic field strengths and rotation rate.

Until relatively recently the planets known to be highly magnetized shared the satisfying property of having their externally observed fields be nearly aligned (within of the order of ten degrees) with the planet's rotation axis, and nearly centered on the planet's body. This property was pleasing again for the reason that the magnetic field generation process is thought to depend on rotation as the main organizing influence on the fluid motion. However, spacecraft observations made near Uranus and Neptune reveal that the magnetic fields of both those objects depart from the simple picture. Models based on the flyby data reveal that both magnetic fields deviate from the center of the planet by some several tenths of the core radius, and that, in each case, the field axis departs from the respective rotation axis by at least several tens of degrees.

In addition to the measurable external, poloidal magnetic fields, it is certain that each planet also has a (possibly much stronger) internal toroidal magnetic field confined largely to highly electrically conducting parts of the planet's interior—where the field is actually generated. The reason for this will become apparent presently.

Only for Earth do we have sufficient data to construct a picture of the magnetic field's temporal behavior. Paleomagnetic studies, making use of the geological record of rock magnetized in Earth's magnetic field reveal that the contemporary terrestrial field is more or less typical—an axially aligned dipolar field, with an intensity that, while not constant, usually only varies by about a factor of two. Remarkably, however, the polarity of the magnetic field is not fixed. Earth's magnetic field changes polarity, suddenly and quickly, at random intervals that average between 100,000 and 200,000 years, though intervals of much shorter and much longer duration also appear in the record. In fact, the distribution of polarity intervals is similar to a Poisson distribution. Briefly, Earth's magnetic field seems to have the aspect of a nonperiodic, bistable oscillator.

In this summary of the properties of known planetary magnetic fields, I have purposely avoided giving a highly detailed and quantitative listing of the various field parameters. Instead I have tried to extract those broad general properties that challenge our understanding at its current state of development.

1.2 THE SUN

A discussion of planetary magnetic fields cannot be complete without reference to the other large cosmic magnetic field generated in the Solar System: the magnetic field of the Sun. If for no other reason, the Sun is both interesting and instructive for what it teaches us about the variety of dynamo magnetic fields that exist in real bodies.

The solar magnetic field suffers two complications as it is observed. Firstly, unlike the planets which hide their field-generation regions beneath thick mantles of poor electrical conductor (Figure 1), the Sun reveals the top of the generation region directly to our observation. Thus much of the simplification imposed on the observable parts of planetary magnetic fields is stripped away when observing the Sun; with the Sun, we look directly into the complication of a turbulently convective, electrically conducting fluid. Secondly, the

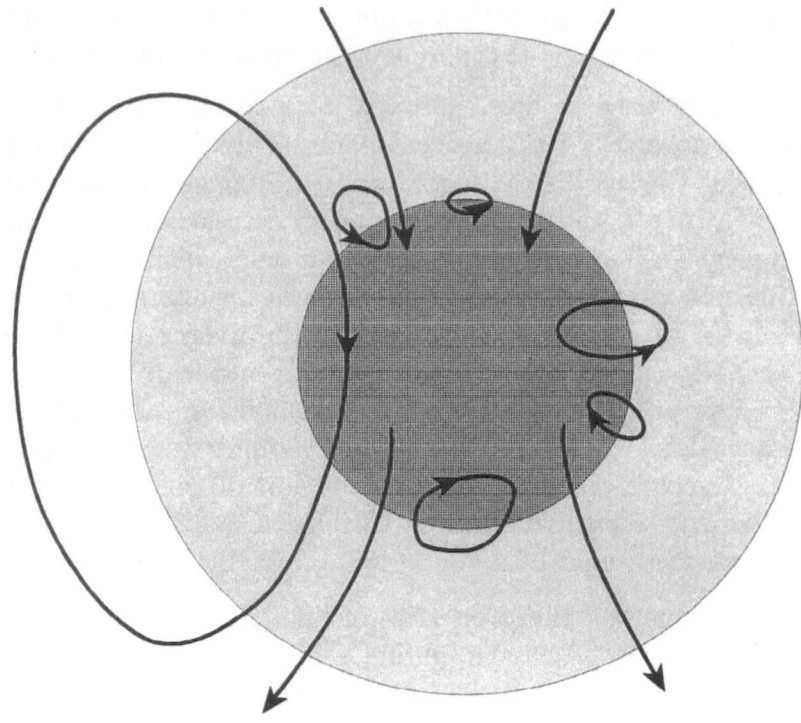

Figure 1. The high order, small scale components of a magnetic field di-
minish rapidly with distance from the center. In planets like Earth, the
dynamo generation region is buried well within the body, so that the field
observed above the surface has a far simpler structure than that which actually exists in the generation region. Other things being equal, the closer
the field-generation region to the measurable surface, the more complicated
will be the structure of the surface field.

Sun is surrounded by a highly electrically conducting corona, which
accelerates into the surrounding space, ultimately to form the solar wind. This expanding gas stretches many of the field lines into
an open configuration, very different from what is observed in the
not-to-distant parts of planetary magnetic fields.

However, abstracting away from those and other complications,
reveals a solar magnetic field that has a generally dipole-symmetry
(even though quite complex) structure. From the perspective of this
article, the most striking aspect of the solar magnetic field is its

temporal behavior. The Sun's "normal" magnetic state seems to be that of a regular oscillation, with an approximately 22-year full period.

1.3 TEMPORAL BEHAVIOR AND SPATIAL CHARACTER

Except in the cases of Earth and Sun, Solar System magnetic fields have been seen only in snapshot. Consequently, we are ignorant of the temporal behavior of the other planetary magnetic fields. This ignorance is particularly sharp in the case of Uranus and Neptune, where the magnetic field configurations—offset so far from their planets' centers and deviating so far from axial alignment—turned out to be so surprising. In these two cases, especially, interpretation of the field structures would be aided by long-term synoptic measurements; however, the timescales daunting for the kinds of *in situ* measurements that are needed.

The distinction between the magnetic fields of Uranus and Neptune and the fields of the other planets may result from differences in their the geometry of the generation regions. Uranus and Neptune may generate their magnetic fields in relatively shallow and thin shells, while the other planets' generation regions are thought to be deeper, largely filled, spherical regions. Such a difference could account for the less organized appearance of the Uranian and Neptunian fields. Indeed, the Sun shows similar levels of structural disorganization.

2. Dynamo Magnetic Field Generation

The most compelling theory—indeed the only viable theory—for the generation of cosmic magnetic fields, such as those observed in the Sun and planets, is the magnetohydrodynamic dynamo. Conceptually, the dynamo problem is easy to state: How does the natural, internal motions of an electrically conducting fluid, in the presence of a magnetic field, generate electrical currents that reinforce the magnetic field, and prevent it from decaying to zero over a long time? The mathematical and logical foundations of the dynamo problem are the magnetohydrodynamic induction equation, easily derivable from Maxwell's equations and Ohm's law,

$$\frac{\partial \mathbf{B}}{\partial t} = \nabla \times (\mathbf{v} \times \mathbf{B}) + \eta \nabla^2 \mathbf{B} \qquad (1)$$

and Newton's laws as embodied in the equation of fluid motion,

$$\rho \frac{\partial \mathbf{v}}{\partial t} + \rho \mathbf{v} \cdot \nabla \mathbf{v} = -\nabla p + \frac{(\nabla \times \mathbf{B}) \times \mathbf{B}}{4\pi} + \cdots, \qquad (2)$$

along with the pertinent constitutive relations and boundary conditions. The magnetic diffusivity, η, is given by $c^2/4\pi\sigma$, where σ is the electrical conductivity. The dynamo problem involves finding solutions with $\mathbf{B} \neq 0$ after a long time.

Taking the dimensional ratio of the first to the second term on the right side of eq. (1) defines the dimensionless magnetic Reynolds number, $\mathcal{R}_m$.

$$\mathcal{R}_m \equiv \frac{vB/\ell}{\eta B/\ell^2} = \frac{v\ell}{\eta}.$$

$\mathcal{R}_m$ measures the extent to which the magnetic field is influenced by the fluid in which it is embedded. When $\mathcal{R}_m < 1$, the second, dissipation term dominates, with the magnetic field lines slipping easily through the fluid. When $\mathcal{R}_m > 1$, the first, convection term dominates, and magnetic field lines are locked effectively to the fluid, and carried along with it.

Similarly, from eq. (1), we can define the magnetic diffusion time, $\tau_m = \ell^2/\eta$. For a metallic sphere a few thousand kilometers across, $\tau_m \sim 10^4$ years. $\mathcal{R}_m$ can be thought of as τ_m/τ_c, where τ_c, the time scale of the convective motions. Thus, in the example given here, convective motions having a length scale of several thousand kilometers, and a time scale substantially shorter than 10^4 years, will significantly influence an embedded magnetic field.

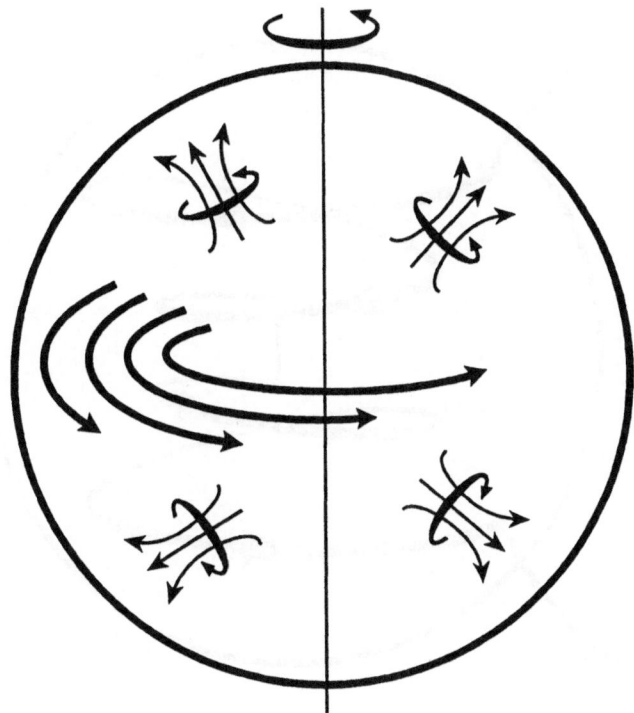

Figure 2.   The fluid motions involved in the basic dynamo process are the large scale differential rotation and the smaller scale helical or cyclonic convection. Both are likely to be present in most cosmic dynamos, but the relative contribution of each is likely to vary from object to object.

Figure 2 depicts the two fluid motions, which, working together, are thought to be most effective generators of magnetic fields in cosmic bodies. Both parts of the motion are intimately related to the bodies' rotation. One part is the regular part of the nonuniform, or differential, rotation, in which, for a spherical body, the angular velocity of the fluid is a function of r, θ, so that, $\Omega = \Omega(r, \theta)$. The other part of the motion is cyclonic, or helical, convection, which ($(\nabla \times \mathbf{v}) \cdot \mathbf{v} \neq 0$) locally.

In its most transparent manifestation, the basic dynamo generation mechanism can be thought of as a two-stage process. The differential rotation winds poloidal magnetic field lines around the axis to generate a toroidal magnetic field, B_ϕ, which may be considerably

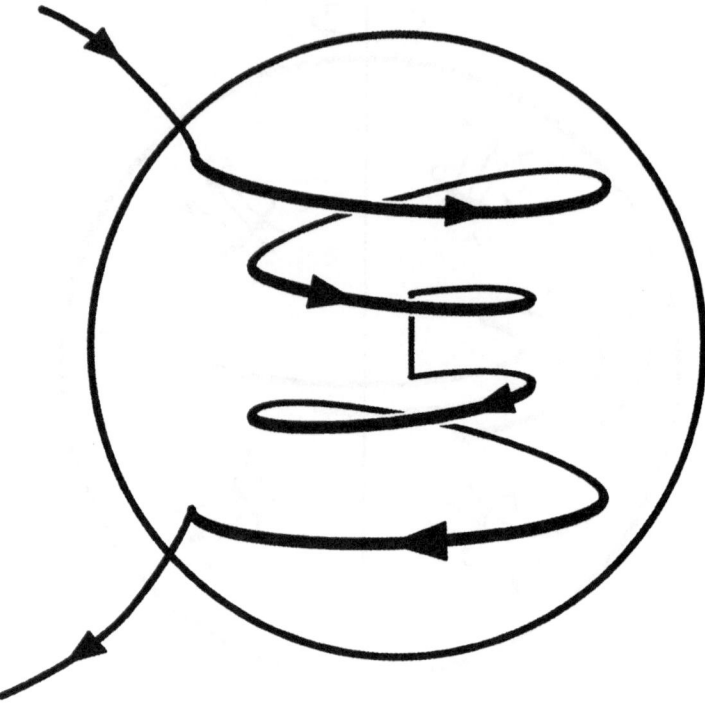

Figure 3.   The differential rotation winds poloidal magnetic field lines around the rotation axis to generate toroidal magnetic field within the dynamo region. Depending on the specific physical regime of the system, the toroidal magnetic field within the dynamo may be much larger than the poloidal field—though it has not directly observable manifestation outside.

stronger than the poloidal, as illustrated in Figure 3. Mathematically, with $v\phi = r\Omega(r)\sin\theta$ eq. (1) yields,

$$\frac{\partial B_\phi}{\partial t} - \eta\left(\nabla^2 - \frac{1}{r^2\sin^2\theta}\right)B_\phi = \left(\frac{\partial v_\theta}{\partial r} - \frac{v_\theta}{r}\right)B_{\rm r} + \cdots$$

$$\equiv \gamma B_{\rm p} \qquad [\approx \frac{v_\phi}{\ell}B_{\rm p}], \tag{3}$$

where the dots indicate additional terms that appear if the angular velocity also a function of θ. The helical convection distorts the

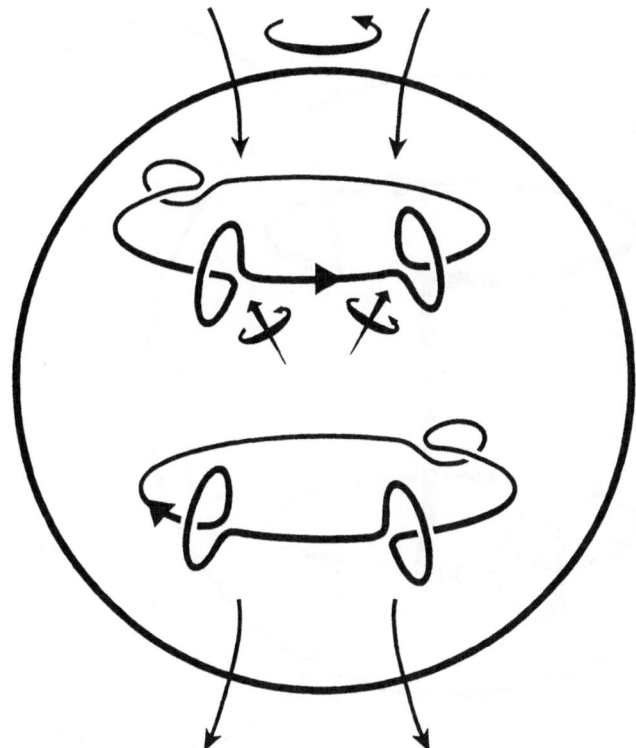

Figure 4. The cyclonic convection distorts the toroidal magnetic field, producing loops of poloidal field.

toroidal magnetic field creating loops of poloidal field, as illustrated in Figure 4. Mathematically,

$$\frac{\partial A_\phi}{\partial t} - \eta \left(\nabla^2 - \frac{1}{r^2 \sin^2 \theta} \right) A_\phi = \Gamma B_\phi \qquad [\approx v_{\text{cyc}} B_\phi]. \qquad (4)$$

Γ and γ are functionals of the fluid velocity field. For a fully convective fluid, $\Gamma \sim v_{\text{cyc}}$, where v_{cyc} is the helical, or cyclonic, part of the convective motion, i.e., that part of the motion corresponding to the screw-like motion turning around the axis of a convective eddy (see Levy 1976 for a simple physical discussion.) We have written

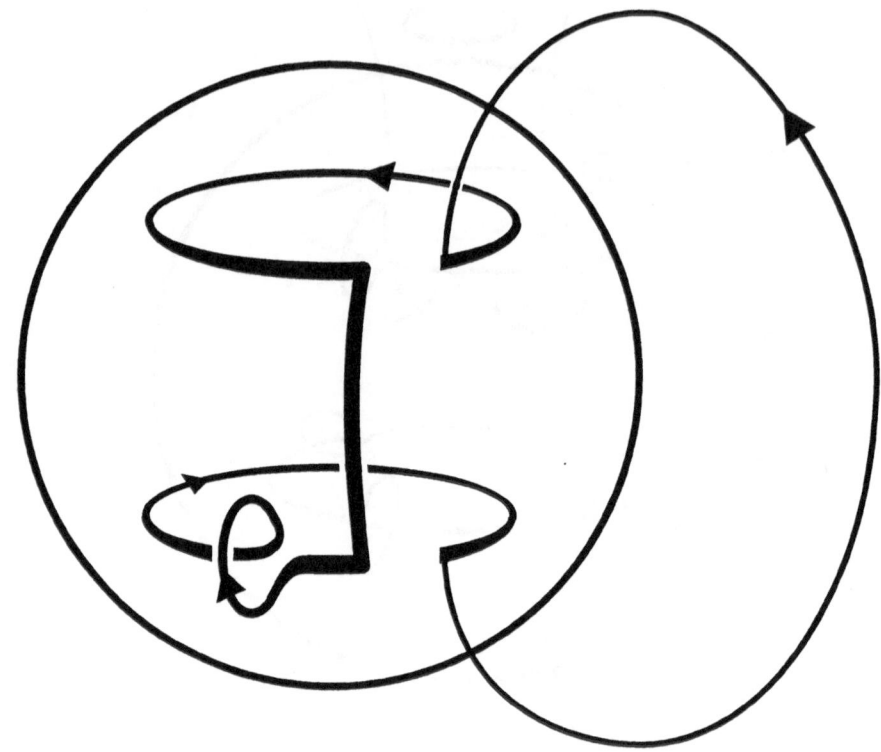

Figure 5. The poloidal loops diffuse outward, ultimately becoming part of the large scale, external magnetic field.

$$\mathbf{B}_\mathrm{p} \equiv (B_\mathrm{r}, B_\theta) = \nabla \times \mathbf{A}_\phi. \tag{5}$$

The small poloidal loops, once having been generated, diffuse outward to become part of the main magnetic field, as illustrated in Figure 5.

This picture and the equations are derived from a set of physical and mathematical ideas and approximations, and will not be discussed further here. The original theory is based on early work by Parker (1955), and has been expanded by much subsequent analysis. The specific form of the right-hand side of equation (4) is based on the approximation of high magnetic Reynolds number in a convective flow with idealized, short-lived, convective eddies (Parker 1955; 1970a,b; 1979). Other, also idealized, derivations lead to essentially the same result (see, example, Steenbeck and Krause 1966; Levy 1978; Parker 1979; Krause and Rädler 1980). In writing equations (3) and (4), we have confined ourselves to only the dominant right-hand-side terms; the reader is encouraged to consult standard review texts on the subject for a more complete discussion (Parker 1979; Moffatt 1978; Krause and Rädler 1980). The dynamo equations are richer than is displayed here, with additional terms that may play a role in field generation under some circumstances. For example, in its full elaboration, eq. (3) also has source terms driven by the convection.

2.1 THE CHARACTER OF THE GENERATED FIELDS

Stationary magnetic fields. It is instructive to examine the underlying structure of this process. First assume stationary magnetic fields in eqs. (3), (4), and (5), so that $\partial/\partial t \equiv 0$, and cast the equations in dimensional form:

$$\eta \frac{\mathbf{B}_\phi}{\ell^2} \sim \frac{v_\phi}{\ell}\mathbf{B}_\mathrm{p},$$

$$\eta \frac{\mathbf{A}_\phi}{\ell^2} \sim v_\mathrm{cyc}\mathbf{B}_\mathrm{p},$$

$$\mathbf{B}_\mathrm{p} \sim \frac{\mathbf{A}_\phi}{\ell}.$$

This homogeneous linear equation set requires, the existence of a solution, with non-zero magnetic field, the consistency relation

$$\mathcal{N} \equiv \frac{v_\phi v_\mathrm{cyc} \ell^2}{\eta^2} = 1.$$

The dimensionless "dynamo number", $\mathcal{N}$ is the product of the two magnetic Reynolds numbers corresponding to the dynamo number for the differential rotation and the dynamo number for the helical convection. The relationship expresses the fact that, if a solution is to exist, the rate at which the magnetic field diffuses away as a result of resistive dissipation must be compensated for by the generation of new magnetic field. Here $\mathcal{N} = 1$ is a criticality condition. For $\mathcal{N} < 1$, the rate of field generation is insufficient to balance dissipation, and the field decays; for $\mathcal{N} > 1$, the rate of field generation exceeds the rate of dissipation, and the field grows. The same general properties are manifested in full, formal solutions of the equations, except that criticality occurs at some value of $\mathcal{N}$ other than unity. Beyond that, the solution set is far richer than this, with a spectrum of stationary, growing, and decaying solutions, as well as oscillatory solutions with stationary, growing, and decaying amplitudes, parameterized by $\mathcal{N}$. These many field states are discussed in a large literature dealing with solution of these (and other, related, dynamo equations).

Oscillatory magnetic fields. A simple analysis due to Parker (1971) can be used to reveal the general character of the oscillating dynamo modes. Rewrite eqs. (3), (4), and (5) respectively in flat, Cartesian coordinates.

$$\frac{\partial B_y}{\partial t} - \eta \nabla^2 B_y = \gamma B_x,$$

$$\frac{\partial A_y}{\partial t} - \eta \nabla^2 A_y = \Gamma B_y,$$

and

$$B_x = -\frac{\partial A_y}{\partial z}.$$

These three equations have solutions in the form of propagating waves, which can be written in the form

$$\begin{pmatrix} A_y \\ B_y \end{pmatrix} = \begin{pmatrix} A \\ B \end{pmatrix} \exp\left(ikz + i\omega t\right).$$

Inserting this *ansatz*, and solving for the resulting dispersion relation, we find:

$$\begin{pmatrix} A_y \\ B_y \end{pmatrix} = \begin{pmatrix} A \\ B \end{pmatrix} \exp\left[ikz \pm i\sqrt{\frac{|\gamma\Gamma k|}{2}}\, t \right] \exp\left[\left(-\eta k^2 + \sqrt{\frac{|\gamma\Gamma k|}{2}} \right) t \right].$$

The basic dynamo modes are given in the form of propagating waves with amplitudes that can either grow, decay or be stationary, as mentioned earlier. Fitting these modes into a finite body—eg., a planet—with applicable boundary conditions, results in an oscillating magnetic field.

For a stationary amplitude oscillating dynamo, it is necessary that the argument of the second exponential function vanishes. Thus, the condition for a stationary amplitude dynamo oscillation in this simple idealization reduces to

$$\frac{\gamma\Gamma k^{-3}}{2\eta^2} = 1,$$

qualitatively similar, as it must be, to the condition for stationary dynamo mode generation in the dimensional analysis given earlier. This stationary-amplitude condition allows us to write the solution in the form

$$\begin{pmatrix} A_y \\ B_y \end{pmatrix} = \begin{pmatrix} A \\ B \end{pmatrix} \exp\left(ikz \pm i\nu_d t \right),$$

where ν_d is the free-decay dissipation rate,

$$\nu_d = \eta k^2.$$

Thus, the characteristic period of the fundamental mode of an oscillating planetary magnetic field should be of the order of the free resistive decay time of the planet. For Earth, this would be of the order of 10^4 years, though we already know that the terrestrial magnetic field is a quasistationary reverser, a rather different behavior from that of a regular oscillator. For the other highly magnetized planets, we do not know whether the dynamos are operating in stationary or oscillating modes. To estimate the periods of those fields were they, in fact, oscillators, would require a more detailed discussion than fits here, including consideration of the possible effects of turbulent convection on the dissipation times (Levy 1976). Large planets, with extensive dynamo regions would be expected to have long oscillation times, probably of the order of 10^3 to 10^4 years. (The relatively short-period, 22-year solar magnetic oscillation is thought to be driven by very rapid mixing in the Sun's turbulent convection zone, where fluid speeds approach a kilometer per second. However, substantial questions persist about the efficacy of such turbulent dissipation in dynamo systems with very high magnetic Reynolds numbers [Vainshtein and Cattaneo 1992]. Lorentz forces may inhibit those very components of the turbulent motion that are responsible for mixing the magnetic field and enhancing dissipation. In planetary dynamos, with their much smaller magnetic Reynolds numbers, turbulent dissipation plays a much smaller role, the same uncertainties do not forcefully arise.) On the other hand, if, for example, Uranus and Neptune were generating their fields in thin, weakly electrically conducting shell dynamos, oscillating field mights be generated with significantly shorter periods. Exploration of the outer solar system in the medium- to long-term future should aim for accurate, time-spaced measurements of the magnetic fields of those planets, in order to ascertain how their peculiar geometries might be related to time variation. Mercury may also be generating its magnetic field in relatively thin, shallow shell.

2.2 THE STRENGTHS OF THE GENERATED FIELDS

Up to this point, we have couched the discussion in terms of the so-called kinematic dynamo, examining the generation of a magnetic field from fluid motions that are asserted without a full treatment of the dynamics giving rise to the motions, and without reference to the magnetic field's back reaction on the fluid. This restricted approach does not allow addressing such questions as the strength of the generated fields—a complicated question that is likely only to yield, in the end, to numerical approaches. However, a few remarks are likely to be worthwhile here. We have already alluded to the fact that the magnetic field enters the dynamical equation (2) directly through the Lorentz force. Thus the dynamically significant quantity is the magnetic field itself, not such secondary aspects as the magnetic moment of a body. Similarly, the other dynamically important influences enter through the buoyancy, pressure and Coriolis forces, etc. Such secondary aspects as the body's total angular momentum are unlikely to play an important role in controlling a planet's magnetic field strength. Because of this, the frequently discussed "magnetic Bode's laws"—which show a linear relationship, on a log-log scale, between planetary magnetic moments and angular momenta—are unlikely to be of any deep significance. To extent that such a relationship seems to apply to the planets, it is likely to be dominated by the fact that both the angular momenta and the magnetic moments of bodies scale as powers of their radii: other things being equal, the ratio of the logs ought to be linear (a point emphasized by K. Runcorn, among others).

While other things are not exactly equal, the strengths of the poloidal magnetic fields within the dynamo regions, the fluid densities, and the magnetic field strengths vary only by factors of a few from one object to the next, among the highly magnetized planets. However, the actual dynamical balance within a dynamo will depend not only on the observable poloidal magnetic field, but also on the toroidal magnetic field, which is hidden from view within the electrically conducting interior. As mentioned earlier, the Coriolis force imposes the organizing influence on the fluid motions that produces effective magnetic field generation. It is plausible that the ultimate dynamical balance within dynamos will be governed by some kind of balance between the Coriolis force ($\sim \rho v \Omega$) and the Lorentz force ($\sim B_p B_\phi / 4\pi\ell$), though that is not the only possibility.

3. Summary and Conclusions

It is compelling to believe that planetary magnetic fields–at least for the highly magnetized planets—are generated by a hydromagnetic dynamo process in the electrically conducting, convecting, fluid interiors of these bodies. Although we currently understand, in general terms, how this process works, the construction of detailed models for individual objects still eludes us. The limitation comes from two directions. We do not have a confident enough knowledge of the structures and motions within most planets to provide a firm foundation on which to build models of magnetic field generation. And, also, our ability to handle the complex dynamical problems— including the full solution of the fluid motion as well as the effects of the magnetic forces—is still at a relatively primitive stage of development.

4. Acknowledgement

This work was supported in part by a grant from the National Aeronautics and Space Administration.

5. References

Krause, F. and Rädler, K.-H. (1980) *Mean-Field Magnetohydrodynamics and Dynamo Theory*, Pergamon Press.

Levy, E.H. (1976) Generation of Planetary Magnetic Fields, *Annual Reviews of Earth and Planetary Science*, **4**, 159–185

Levy, E.H. (1978) Magnetic field generation at high magnetic Reynolds number, *Ap. J.*, **220**, 325–329.

Moffatt, H.K. (1978) *Magnetic field generation in electrically conducting fluids*, Cambridge University Press.

Parker, E.N. (1955) Hydromagnetic dynamo models, *Ap. J.*, **122**, 293–314.

Parker, E.N. (1970a) The origin of magnetic fields, *Ap. J.*, **160**, 383–404.

Parker, E.N. (1970b) The generation of magnetic fields in astrophysical bodies. I. The dynamo equations, *Ap. J.*, **162**, 665–673.

Parker, E.N. (1971) The generation of magnetic fields in astrophysical bodies. I. The solar and terrestrial fields, *Ap. J.*, **164**, 491–509.

Parker, E.N. (1979) *Cosmical Magnetic Fields*, Oxford University Press.

Steenbeck, M. and Krause, F. (1966) *Z. Naturforsch. A*, **21**, 369.

Vainshtein, S.I. and Cattaneo, F. (1992) Nonlinear restrictions on dynamo action, *Ap. J.*, **393**, 165–171.

PLANETARY MAGNETOSPHERES

MICHAEL SCHULZ

Space Sciences Department, Lockheed Palo Alto Research Laboratories
Palo Alto, California, U.S.A. 94304

Abstract. Two decades of *in situ* planetary exploration with fly-by missions have revealed a rich variety of magnetospheric configurations and dynamical phenomena, some anticipated and some remarkably surprising. These discoveries have set the stage for further exploration of planetary magnetospheres by orbiting spacecraft.

1. Introduction

The usual configuration of a planetary magnetosphere results from dayside compression and nightside extension of an intrinsic planetary magnetic field by the solar wind. The planetary magnetic field is thus confined (in first approximation) behind an elongated boundary known as the magnetopause, whose position is determined largely by the requirement of kinetic pressure balance between the magnetic field and the solar wind (cf. Figure 1). Typical subsolar radii are about $1.5\ R_M$ for Mercury's magnetopause, 10 R_E for Earth's, 50–100 R_J for Jupiter's, 20 R_S for Saturn's, 17 R_U for that of Uranus, and 26 R_N for Neptune's. The solar wind is a supersonic and super-Alfvénic plasma (mainly hydrogenic) whose velocity **u** can vary (with heliomagnetic latitude and time) from about 300 km/s to about 800 km/s (350 km/s being typical). The plasma density N at heliocentric distance $R = 1$ AU is likewise variable ($\sim$ 1–30 cm^{-3}, with 8 cm^{-3} being typical). Kinetic pressure balance between the planetary magnetic field and the solar wind leads to a first estimate $b \approx (8\pi N \langle m \rangle u^2 / \mu_P^2)^{1/6}$ for the dayside (subsolar) radius b, where μ_P is the strength of the planetary dipole moment. Each planetary magnetosphere is preceded by an upstream bow shock, roughly hyperboloidal in shape and with a subsolar radius ~ 1.3 times the subsolar radius of the magnetopause.

The **B** field in the distant nightside magnetosphere (at distances $\gtrsim b$) is typically "stretched" by a cross-magnetospheric current sheet so as to form a long magnetotail. The current sheet divides the tail into two "lobes" of open magnetic field lines with opposite magnetic polarities. The current-sheet configuration depends mostly on the angle ψ between the dipole moment and solar-wind velocity (cf. Figure 2, which simulates Neptune's magnetosphere). [Selected field lines in Figure 2 emanate from a dipole-centered sphere of radius $r = a = b/26$) at 2° intervals of magnetic latitude λ (viz.,

Earth, Moon, and Planets 67: 161–173, 1995.
© 1995 *Kluwer Academic Publishers.*

±90°, ±88°, ±86°, . . ., 0°) and 15° intervals of magnetic local time φ. Coordinates are ξ ≡ distance downstream from the planet, η ≡ distance from "noon-midnight" meridional plane, and ζ ≡ distance from the plane that perpendicularly intersects this meridional plane along the tail axis.] There is a model-dependent critical angle ψ* ≈ 25°, such that for ψ < ψ* and for ψ > 180° − ψ*, the nightside current sheet closes on itself within the tail rather than connecting with the magnetopause current. In this case the summer tail lobe becomes annular in shape and engulfs the winter lobe. This was the configuration of Neptune's magnetosphere as Voyager 2 entered through the polar-cleft region (Day 236.8, 1989: Connerney *et al.*, 1991). The angle ψ was about 20° at that time (R. P. Lepping, personal communication, August 1993). Uranus' magnetosphere will be able to achieve the same topology early next century.

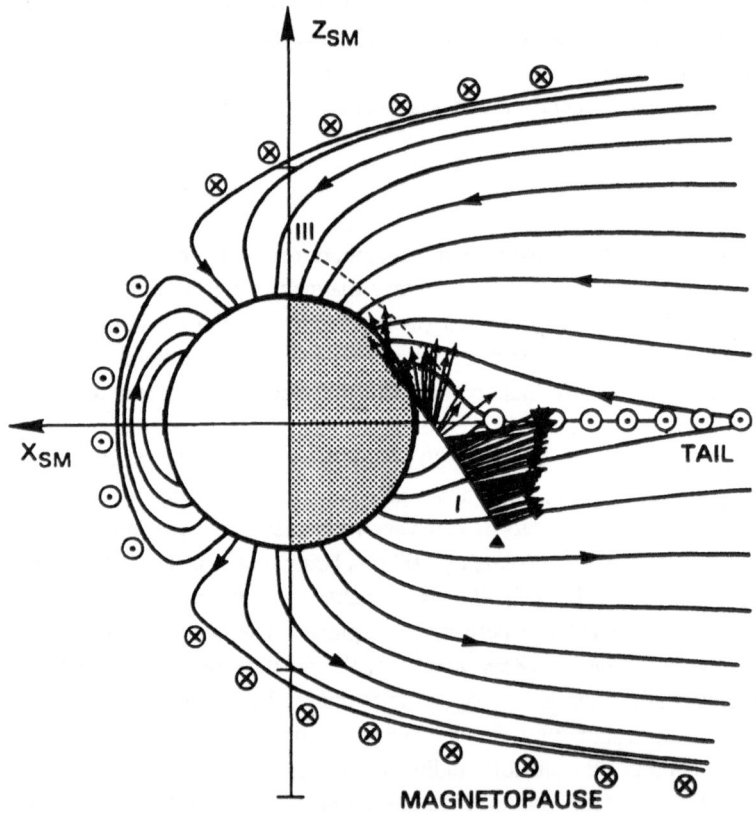

Figure 1. Field-line configuration in noon-midnight meridional plane of Mercury's magnetosphere (Connerney and Ness, 1988; after Whang and Ness, 1975; Jackson and Beard, 1977), based on observations by Ness *et al.* (1975). Also shown: Projections of Mercury I (March 1974) and Mercury III (March 1975) encounter trajectories of Mariner 10 and of Mercury I magnetic-field data.

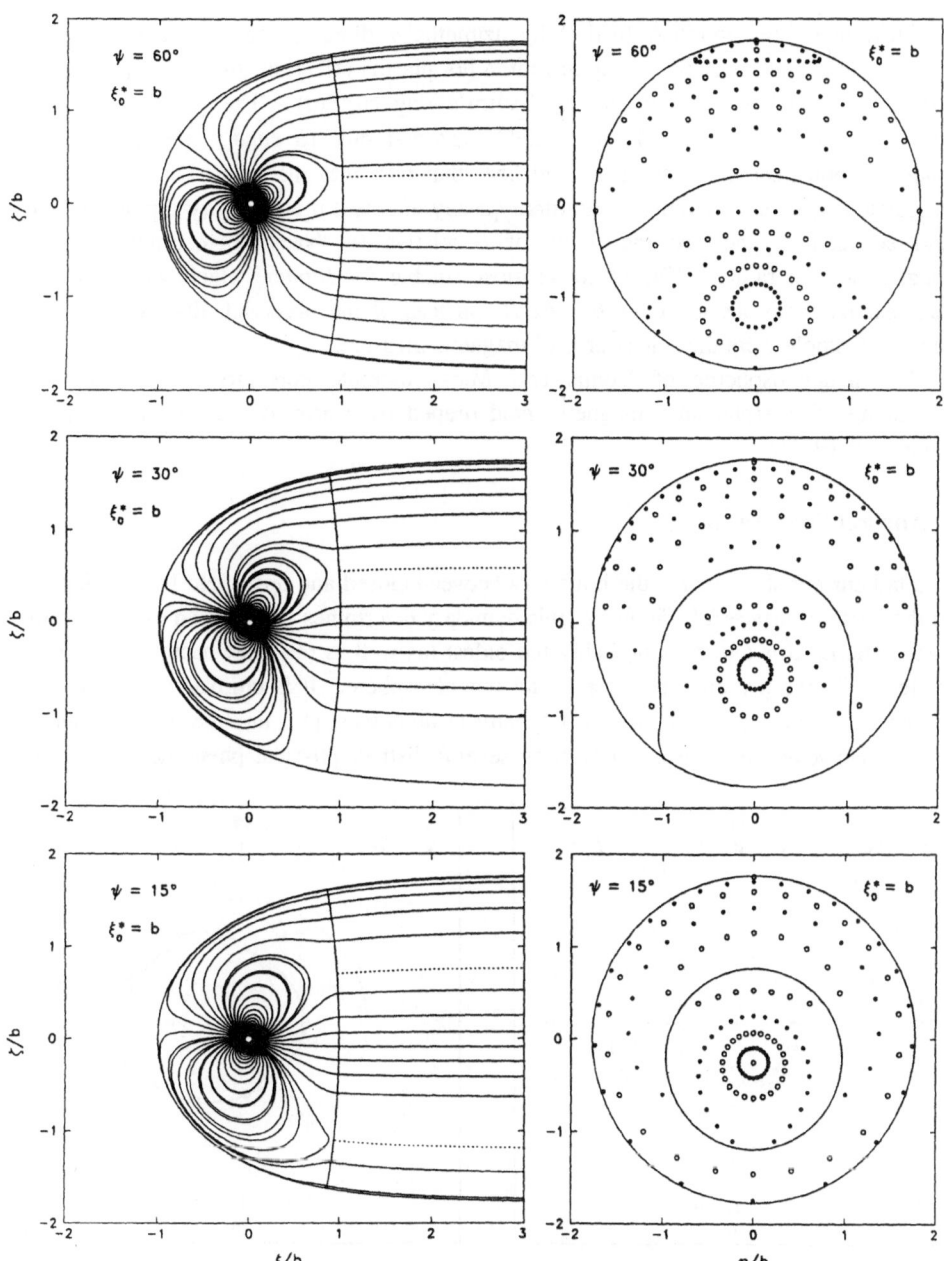

Figure 2. Source-surface model of Neptune's magnetosphere (Schulz *et al.*, 1995). Left-hand panels show representative field lines in "noon-midnight" meridional plane. Right-hand panels show intersections of representative field lines with cross section of distant magnetotail. Parameter b (= 26 R_N) denotes distance from point dipole to subsolar point on magnetopause.

Field lines are stretched further by azimuthally directed internal magnetospheric currents (e.g., by the Earth's ring current, a hot-plasma effect; by Jupiter's centrifugally driven magnetodisk current; and by Saturn's ring current). The presence of such currents tends to "open" additional field lines and thus to enlarge the area of the magnetic polar cap (consisting of field lines that map into the tail). Dipole moments associated with such azimuthal currents typically augment the planetary dipole moment and so tend to increase the value of b relative to "vacuum" estimates of the magnetospheric radius. The effect is most notable for Jupiter, which would have a subsolar magnetospheric radius $b \approx 42\ R_J$ (instead of the observed 50–100 R_J) if the planetary dipole were the sole source of magnetic pressure.

The magnetospheres of Venus and Mars resemble cometary magnetospheres, consisting of interplanetary magnetic field draped over around the ionosphere (e.g., Luhmann, 1991).

2. Auroral Phenomena

The tail current sheet marks the boundary between closed and open field lines. Mapped to the Earth's surface (cf. Figure 3), this boundary determines the position of the auroral oval. The Earth's aurora is probably the oldest (as well as the most beautiful) piece of evidence for the presence of a terrestrial magnetosphere. It is ultimately an electrical discharge, driven by magnetospheric dynamical processes, in a mixture of atmospheric gases. However, the aurora comprises several distinct physical phenomena (discrete

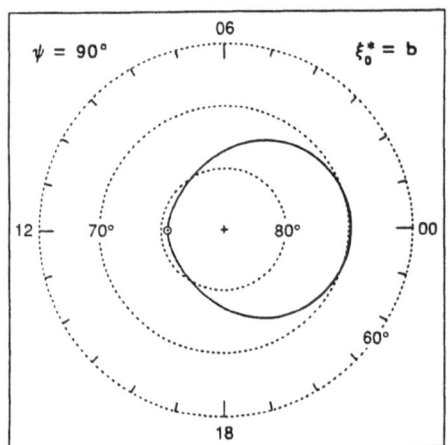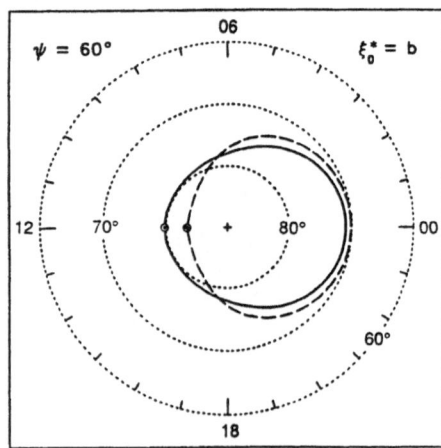

Figure 3. Boundaries between closed and open field lines, as obtained from source-surface model of magnetosphere for selected values of ψ and mapped to the Earth's surface ($r = a = b/10$). Solid curve pertains to winter hemisphere, dashed curve to summer hemisphere, for $\psi \neq 90°$ (Schulz and McNab, 1995). Sun symbols ($\odot$) indicate field-line mappings from dayside neutral points.

arcs, as well as diffuse and patchy forms) even at Earth (e.g., Akasofu, 1976). Discrete arcs seem to be associated with sharp plasma boundaries along which field-aligned (parallel-to-**B**) electric currents (known as Birkeland currents) and parallel electric fields are directed. The diffuse aurora seems to be a consequence of the pitch-angle scattering of plasma-sheet and ring-current electrons by waves in the magnetospheric plasma, and patchy forms seem to result from modulations caused by larger-scale plasma instabilities in the equatorial magnetosphere.

Auroral phenomena on other planets seem to involve still other physical processes. Jupiter's decametric (DAM) radio emission, for example, appears to result from a field-aligned current driven by the rotation of Jupiter's magnetosphere past the electrically conducting Galilean satellite Io. The emission seems to come from Jupiter's ionosphere at the foot of Io's flux tube, which is warped about 20° in longitude by the field-aligned current (Piddington and Drake, 1968; Goldreich and Lynden-Bell, 1969). The emission mechanism is believed to entail an electromagnetic plasma instability caused by interaction of the current with the Jovian ionosphere (Goldstein and Goertz, 1983), much as auroral kilometric radiation (AKR) from the Earth's upper ionosphere seems to involve an momentum-space instability (Wu and Lee, 1979) of the electron distribution that carries terrestrial Birkeland currents. The Jovian aurora at other longitudes seems to be associated with the Io plasma torus, which consists largely of magnetospherically ionized sulfur from Io's volcanoes. Uranus and Neptune likewise show auroral emissions from latitudes well equatorward of the boundary between closed and open field lines (Farrell et al., 1991; Herbert and Sandel, 1994).

3. Thermal Plasma

The distribution of thermal plasma in a planetary magnetosphere (cf. Belcher et al., 1990) is influenced by gravity, by centrifugal forces, and by large-scale electric fields **E** perpendicular to **B**. Plasma density in the terrestrial magnetosphere shows a sharp discontinuity across **B** at the plasmapause, which is associated with the "last closed equipotential" of **E** (Nishida, 1966). The electric field in this case is the sum of fields that describe magnetospheric corotation and magnetospheric convection. Since convection on closed field lines carries plasma toward the Sun, this superposition typically produces stagnation in the flow at a field line near the dusk meridian (see Figure 4, left panel), and this stagnation line generates a separatrix between closed and open equipotential surfaces.

Convection at Jupiter seems too weak to produce such a stagnation line, and in this case the last closed equipotential surface is one that barely grazes the magnetopause, perhaps along the dusk meridian (Brice and Ioannidis, 1970; cf. Figure 4, right panel). However, Jovian plasma (much of which comes from Io) in the middle and outer magnetosphere is largely confined (e.g., Belcher, 1983) to a nearly equatorial annulus (the magnetodisk) as a consequence of centrifugal forces projected along **B** (Siscoe,

1977). The component of centrifugal force normal to **B** leads to an azimuthal drift current which distorts the magnetic field (see Figure 5) in a way that sharpens the equatorial confinement of Jovian plasma. Such a plasma distribution is unstable against centrifugally driven interchange (Richardson *et al.*, 1980; Siscoe and Summers, 1981), of which the consequence is a net-outward eddy diffusion of magnetodisk plasma continually supplied by Io.

Approximation of the plasmapause as the "last" closed equipotential surface is grossly inappropriate for the magnetospheres of Uranus and Neptune, whose dipoles are tilted so strongly (by 59° and 47°, respectively) that planetary rotation generates mostly induced **E** fields. In such magnetospheres it has been useful (Selesnick, 1988, 1990) to view the problem of cold-plasma drifts in the reference frame of the rotating dipole, from which perspective the convection electric field is strongly modulated in time. The result is an unavoidable transport of cold plasma, which tends to spiral either outward or inward rather than retrace its trajectory as in Figure 4. This transport precludes the formation of a sharp plasmapause in the magnetosphere of a planet such as Uranus or Neptune.

Of course, planets such as Earth and Jupiter (whose dipoles are inclined ~10° relative to the respective rotation axes) are affected to some extent by the same intrinsic transport (Selesnick and Richardson, 1986), but the time-scale for such transport is so much longer in these cases that the idea of a "last" closed equipotential surface remains a reasonably good approximation. Saturn offers the best example of a time-independent

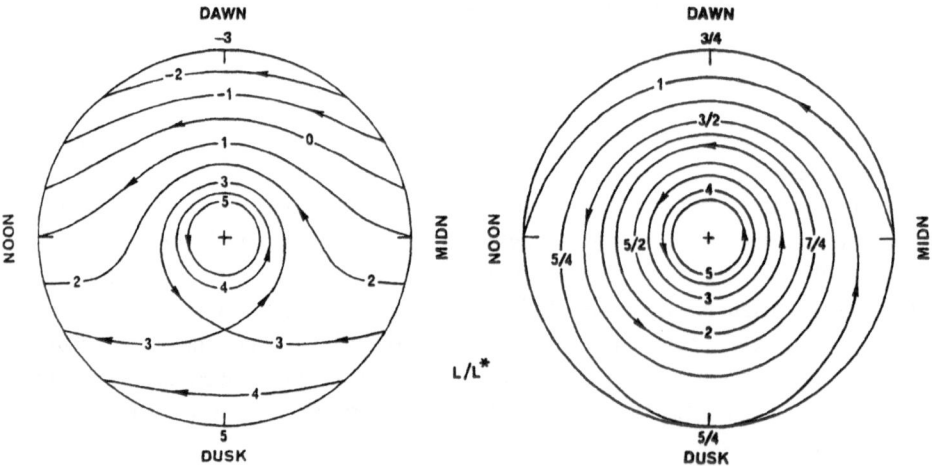

Figure 4. Cold-plasma drift trajectories (Schulz, 1991) in the equatorial plane of an idealized magnetospheric model (Hill and Rassbach, 1975; Volland, 1975; Schulz, 1976). The boundary circle ($L = L^*$) is a neutral line, which means that $B \equiv |\mathbf{B}| = 0$ there. Contour labels represent normalized values of electrostatic potential. Earth-like drift configuration in left-hand panel corresponds to convection electric field 16 times as strong as in right-hand panel. (Weaker convection leads to Saturn-like or Jupiter-like drift configuration.)

magnetospheric configuration (and thus of a purely electrostatic corotation electric field), as its dipole is almost perfectly aligned with the rotation axis (Connerney *et al.*, 1982). However, the expected convection electric field is too weak to stagnate the plasma flow anywhere in Saturn's magnetosphere, and so the last closed equipotential would be the one that grazes the magnetopause or neutral sheet (Siscoe, 1979). The inclination of Mercury's magnetic dipole is not well established but is assumed to be moderate, perhaps 17° (Connerney and Ness, 1988). Even so, Mercury rotates so slowly that magnetospheric convection should dominate magnetospheric rotation at the Sun's innermost planet (Russell *et al.*, 1988).

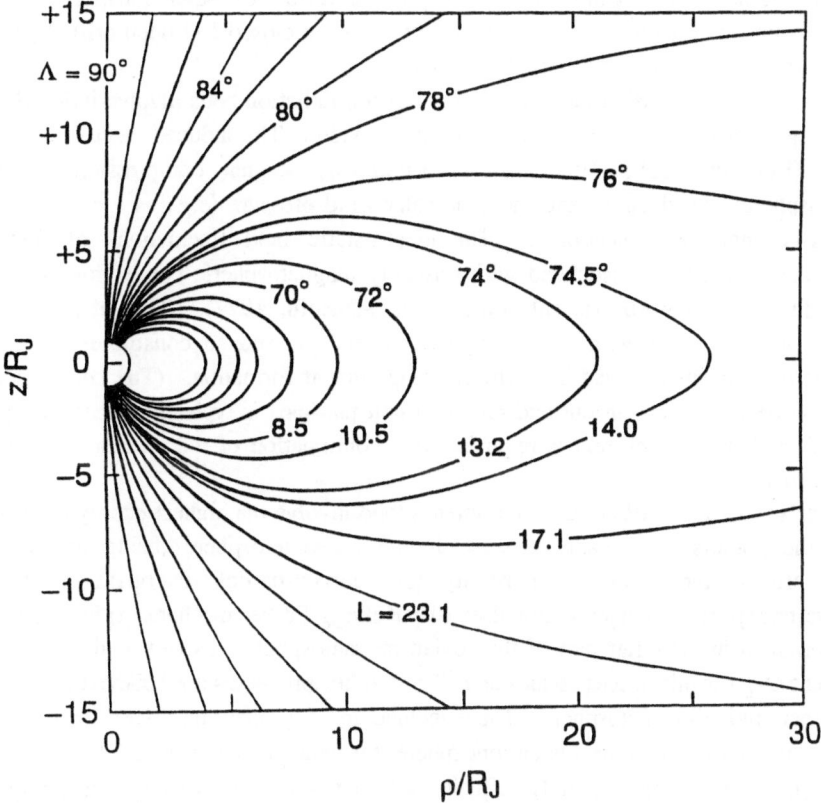

Figure 5. Configuration of field lines (as influenced by magnetodisk current) in Jupiter's middle magnetosphere, according to model of Connerney *et al.* (1981). Value of Λ specifies magnetic latitude at which a field line intersects the planetary surface. Corresponding value of L ($\equiv \sec^2\Lambda$) indicates what would have been the equatorial crossing distance ρ/R_J for the unstretched (dipolar) field line in absence of magnetodisk (composite figure assembled by Schulz *et al.*, 1993).

4. Energetic Particles

Planetary radiation belts consist of relativistic electrons and similarly energetic ions. Although radioactive decay plays some role in the population of radiation belts with energetic particles (Singer, 1958; Blake and Schulz, 1980), the main mechanism for belt formation and particle energization is a large-scale eddy-like diffusion driven by fluctuations is magnetospheric E and B fields. Such fluctuations are typically slow enough to conserve the first two adiabatic invariants (called M and J) of charged-particle motion while violating the third invariant (Φ). Different physical processes are respectively dominant for different particle populations and different planets, but a recurring theme is the net inward (but nevertheless diffusive) radial transport from an external boundary (identified with the transition between closed and open drift shells) at which particles are introduced from the outside.

The main radial-diffusion mechanisms for Earth's radiation belts involve induced and fluctuating electrostatic fields of magnetospheric extent. The induced electric fields of interest (Kellogg, 1959; Parker, 1960) accompany sudden compressions of the magnetosphere by sudden enhancements in solar-wind pressure (e.g., as a consequence of solar-coronal mass ejections). The electrostatic field fluctuations of interest (Fälthammar, 1965) are associated with unsteady magnetospheric convection, such as occurs during magnetospheric substorms (e.g., Akasofu, 1977). The latter process increases in importance with decreasing particle energy, and so constitutes the main transport mechanism involved in stormtime ring-current formation. (The conventional distinction between ring-current and radiation-belt particles is typically placed at about 200 keV, and the ring current is regarded as a continuation of the plasma sheet onto closed drift shells.)

Jupiter's radiation belts are the most intense (and involve the highest-energy particles) among the planets, presumably because Jupiter's magnetosphere (being by far the largest) affords the greatest opportunity for charged-particle energization within. Diffusive transport via magnetic impulses (the Kellogg-Parker mechanism) is presumed to be dominant in the outer part of the Jovian magnetosphere. However, the diffusion coefficient D_{LL} for this process scales as L^{10}, and other processes are believed to prevail in the inner Jovian magnetosphere. These include an eddy diffusion driven by unsteady atmospheric circulation in the Jovian ionosphere, for which D_{LL} scales as L^{10} (Brice and McDonough, 1973), and the eddy diffusion associated with interchange instability of plasma in the Jovian magnetodisk (Siscoe and Summers, 1981). (Regardless of its origin, an electric field interacts with charged particles of any energy and thus contributes something to their transport. Whether a given process is important for a specified particle population is a different question, to be decided by the strength of this process relative to other processes for those particles.)

An important consideration for the radiation belts of Jupiter and the other giant planets is the presence of moons and rings inside the magnetospheres. These obstacles tend to absorb particles from the radiation belts, which would otherwise be considerably more intense (Mead and Hess, 1973). The consequences for Saturn's radiation belts are illustrated by the absorption signatures identified in Figure 6 (Simpson *et al.*, 1980). The mathematical description of particle absorption by satellites and rings is a subject of continuing interest and incremental development (e.g., Paonessa and Cheng, 1985; Paranicas and Cheng, 1991), mainly because of important geometrical complications. Wave-particle interactions and synchrotron emission (see below) are likewise important loss mechanisms for particles trapped in magnetospheric radiation belts.

5. Radio Emissions

The plasma environment of a planetary magnetosphere is susceptible to a great variety of wave-generating instabilities (e.g., Barbosa and Coroniti, 1976; Kurth and Gurnett, 1991), and lightning is a known source of electromagnetic (whistler-mode) waves in several magnetospheres. Some of the waves thus generated are guided so as to remain inside the magnetosphere (e.g., Gurnett and Scarf, 1983; Moses and Coroniti, 1991). Others waves would encounter stop-bands, resonances, or cut-offs (e.g., Allis *et al.*, 1963) *en route* to an external observer. All these phenomena are impediments to radio

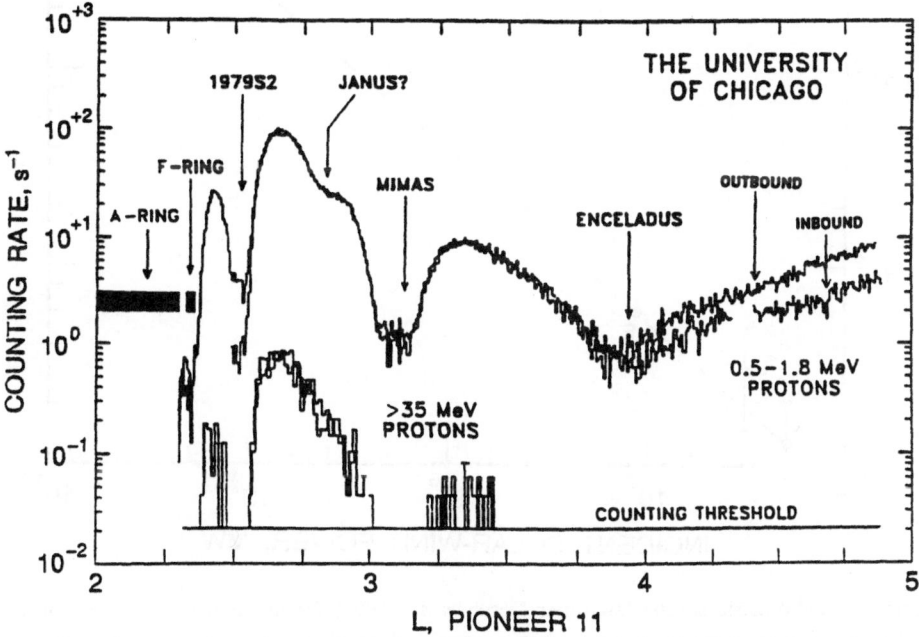

Figure 6. Overlay of inbound and outbound proton-flux profiles from Pioneer-11 encounter (September 1979) with Saturn, showing absorption signatures associated with various rings and moons (Simpson *et al.*, 1980).

emission *from* a planetary magnetosphere. Radio waves are nevertheless emitted from most planetary magnetospheres. Auroral kilometric radiation (AKR) and Jovian decametric radiation (DAM) are important examples and have been mentioned above. These have analogues at several other planets (e.g., Landreiter *et al.*, 1991).

Incoherent synchrotron radiation from radiation-belt electrons constitutes a further radio source. This is the phenomenon responsible for Jovian decimetric radiation (DIM), which appears to come from a pair of roughly elliptical sources centered equatorially about 0.9 R_J above the eastern and western limbs of the planet (Berge, 1966). The apparent source shape is (of course) an illusion resulting from integration along lines of sight through the electron radiation belt, which constitutes an "optically thin" but toroidally distributed medium.

The total low-frequency radio power emitted by a planet shows a remarkably consistent variation (cf. Figure 7) with the solar-wind power incident on the planetary magnetosphere (Desch *et al.*, 1991), as if each planetary magnetosphere somehow finds

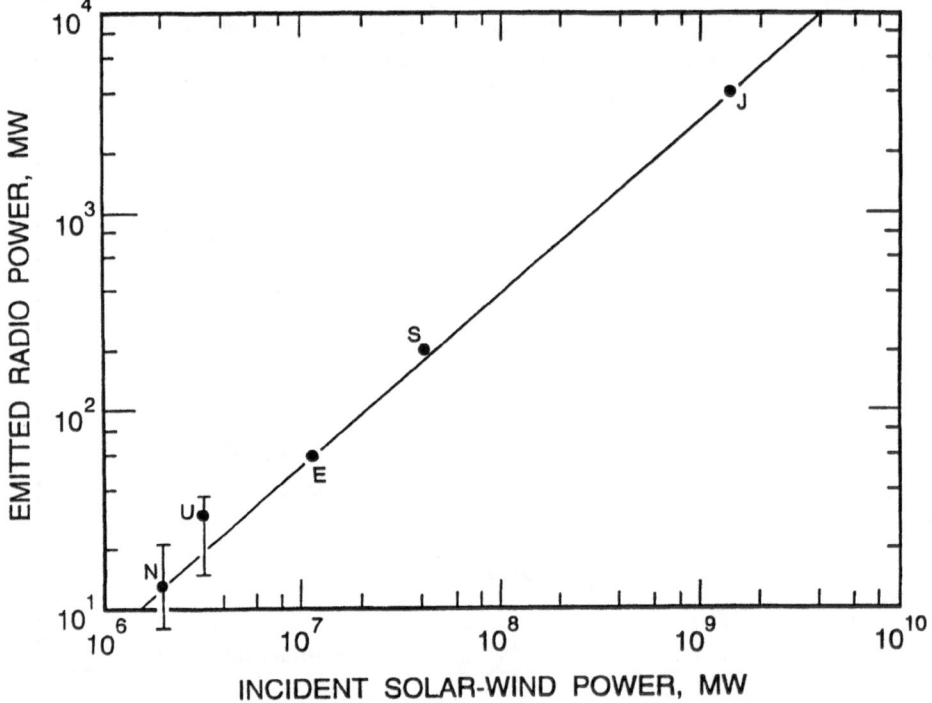

Figure 7. Radiometric scaling law (after Desch *et al.*, 1991) for low-frequency radio power emitted by a planetary magnetosphere, given as empirical function of solar-wind power incident on magnetospheric cross section (modeled as in Figure 2). Straight line (with slope 0.87) corresponds to emission of a fraction $5.26 \times 10^{-6}(P_{inc}^{E} / P_{inc})^{0.13}$ of the incident power P_{inc} in the form of radio waves. Superscript E denotes Earth. Data point for Neptune was generously supplied by M. D. Desch (personal communication, 1994).

a way to convert a small fraction (ranging from 6.5×10^{-6} for Neptune to 2.8×10^{-6} for Jupiter) of the incident solar-wind energy into radio waves of various types. (About 10%–30% of the incident power actually enters the Earth's magnetosphere. This fraction is expected to vary, along with "reconnection efficiency," from planet to planet.)

6. Perspective

While every planetary magnetosphere is different in some way from all the others, all share a dependence on the same basic principles and many have at least some features in common. A broader perspective on magnetospheric physics can be attained by looking beyond the planets themselves. For example, Luhmann (1979) has remarked on the similarity of Jupiter's decimetric emission pattern to the paired radio sources associated with galaxies. Mullan and Schatten (1979) have remarked on the ability of the Sun's magnetic arcade to trap energetic particles in much the way that planetary magnetic fields contain radiation belts. In a very meaningful sense, all known planetary magnetospheres reside within (and interact with) the tail region of the Sun's magnetosphere.

The comparative study of planetary magnetospheres is an intellectual challenge in its own right. Beyond this, however, the dynamical principles elucidated through the study of magnetospheres in general contribute significantly to our understanding of the Earth's space environment. The past two decades of *in situ* planetary exploration have involved (except for Earth and Venus) mainly fly-by missions. Although inherently limited in spatial and temporal coverage, such missions have revealed a rich variety of magneto-spheric configurations and dynamical phenomena. Each planetary magnetosphere encountered has yielded some remarkable surprises, and the discoveries made so far have set the stage for further exploration of planetary magnetospheres by spacecraft that actually orbit the respective planets.

7. Acknowledgments

Preparation of the manuscript for this paper was partially supported by the Independent Research and Development (IR&D) program of the Lockheed Missiles and Space Company. The author thanks M. T. Chahine and N. F. Ness for arranging the invitation, also S. Gulkis and M. D. Desch for their help in connection with Figure 7.

8. References*

Akasofu, S.-I., 1976: Recent progress in studies of DMSP auroral photographs, *Space Sci. Rev.*, **19**, 169–215.
Akasofu, S.-I., 1977: *Physics of Magnetospheric Substorms*, Reidel, Dordrecht.
Allis, W. P., Buchsbaum, S. J., and Bers, A., 1963: *Waves in Anisotropic Plasmas*, M.I.T. Press, Cambridge, MA.

Barbosa, D. D., and Coroniti, F. V., 1976: Relativistic electrons and whistlers in Jupiter's magneto-sphere, *J. Geophys. Res.*, **81**, 4531–4536.

Belcher, J. W., 1983: The low-energy plasma in the Jovian magnetosphere, in A. J. Dessler (ed.), *Physics of the Jovian Magnetosphere*, ch. 3., pp. 68–105, Cambridge Univ. Press, Cambridge, UK.

Belcher, J. W., McNutt, R. L., Jr., and Richardson, J. D., 1990: Thermal plasma in outer planet magnetospheres, *Adv. Space Res.*, **10**, (1)5–(1)13.

Berge, G. L., 1966: An interferometric study of Jupiter's decimetric radio emission, *Astrophys. J.*, **146**, 767–798.

Blake, J. B., and Schulz, M., 1980: The satellites of Jupiter as a source of very energetic particles, *Icarus*, **44**, 367–374.

Brice, N. M., and Ioannidis, G. A., 1970: The magnetospheres of Jupiter and Earth, *Icarus*, **13**, 173–183.

Brice, N. M., and McDonough, T. R., Jupiter's radiation belts, 1973: *Icarus*, **18**, 206–219.

Connerney, J. E. P., and Ness, N. F., 1988: Mercury's magnetic field and interior, in F. Vilas, C. R. Chapman, and M. S. Matthews (eds.), *Mercury*, pp. 494–513, Univ. of Ariz. Press, Tucson.

Connerney, J. E. P., Acuña, M. H., and Ness, N. F., 1981: Modeling the Jovian current sheet and inner magnetosphere, *J. Geophys. Res.*, **86**, 8370–8384.

Connerney, J. E. P., Acuña, M. H., and Ness, N. F., 1991: The magnetic field of Neptune, *J. Geophys. Res.*, **96**, Suppl., 19023–19042.

Connerney, J. E. P., Ness, N. F., and Acuña, M. H., 1982: Zonal harmonic model of Saturn's magnetic field for Voyager 1 and 2 observations, *Nature*, **298**, 44–46.

Desch, M. D., Kaiser, M. L., Zarka, P., Lecacheux, A., LeBlanc, Y., Aubier, M., and Ortega-Molina, A., 1991: Uranus as a radio source, in J. T. Bergstralh, E. D. Miner, and M. S. Matthews (eds.), *Uranus*, pp. 894–925, Univ. of Ariz. Press, Tucson.

Fälthammar, C.-G., 1965: Effects of time-dependent electric fields on geomagnetically trapped radiation, *J. Geophys. Res.*, **70**, 2503–2516.

Farrell, W. M., Desch, M. D., Kaiser, M. L., and Calvert, W., 1991: Evidence of auroral plasma cavities at Uranus and Neptune from radio burst observations, *J. Geophys. Res.*, **96**, Suppl., 19049–19059.

Goldreich, P., and Lynden-Bell, D., 1969: Io, a Jovian unipolar inductor, *Astrophys. J.*, **156**, 59–78.

Goldstein, M. L., and Goertz, C. K., 1983: Theories of radio emissions and plasma waves, in A. J. Dessler (ed.), *Physics of the Jovian Magnetosphere*, ch. 9., pp. 317–352, Cambridge Univ. Press, Cambridge, UK.

Gurnett, D. A., and Scarf, F. L., 1983: Plasma waves in the Jovian magnetosphere, in A. J. Dessler (ed.), *Physics of the Jovian Magnetosphere*, ch. 8., pp. 285–316, Cambridge Univ. Press, Cambridge, UK.

Herbert, F., and Sandel, B. R., 1994: The Uranian aurora and its relationship to the magnetosphere, *J. Geophys. Res.*, **99**, 4143–4160.

Hill, T. W., and Rassbach, M. E., 1975: Interplanetary magnetic field direction and the configuration of the day side magnetosphere, *J. Geophys. Res.*, **80**, 1–6.

Jackson, D. J., and Beard, D. B., 1977: The magnetic field of Mercury, *J. Geophys. Res.*, **82**, 2828–2836.

Kellogg, P. J., 1959: Van Allen radiation of solar origin, *Nature*, **183**, 1295–1297.

Kurth, W. S., and Gurnett, D. A., 1991: Plasma waves in planetary magnetospheres, *J. Geophys. Res.*, **96**, Suppl., 18977–18991.

Landreiter, H. P., Leblanc, Y., Rabl, G. K. F., and Rucker, H. O., 1991: Emission characteristics and source location of the smooth Neptunian kilometric radiation, *J. Geophys. Res.*, **96**, Suppl., 19101–19110.

Luhmann, J. G., 1979: Galactic radiation belts, *Nature*, **282**, 386–388.

Luhmann, J. G., 1991: The solar wind interaction with Venus and Mars: Cometary analogies and contrasts, in *Cometary Plasma Processes*, A. D. Johnstone (ed.), pp. 5–16, Geophys. Monogr. 61, Am. Geophys. Union, Washington, DC.

Mead, G. D., and Hess, W. N., 1973: Jupiter's radiation belts and the sweeping effect of its satellites, *J. Geophys. Res.*, **78**, 2793–2811.

Moses, S. L., and Coroniti, F. V., 1991: A mysterious plasma wave emission and the determination of plasma densities in Neptune's inner magnetosphere, *J. Geophys. Res.*, **96**, Suppl., 19013–19021.

Mullan, D. J., and Schatten, K. H., 1979: Motion of solar cosmic rays in the coronal magnetic field, *Solar Phys.*, **62**, 153–177.

Ness, N. F., Behannon, K. W., Lepping, R. P., and Beard, D. B., 1975: The magnetic field of Mercury, 1, *J. Geophys. Res.*, **80**, 2708–2716.

Nishida, A., 1966: Formation of plasmapause, or magnetospheric plasma knee, by the combined action of magnetospheric convection and plasma escape from the tail, *J. Geophys. Res.*, **71**, 5669–5679.

Paonessa, M., and Cheng, A. F., 1985: A theory of satellite sweeping, *J. Geophys. Res.*, **90**, 3428–3434.

Paranicas, C. P., and Cheng, A. F., 1991: Theory of ring sweeping of energetic particles, *J. Geophys. Res.*, **96**, Suppl., 19123–19129.

Parker, E. N., 1960: Geomagnetic fluctuations and the form of the outer zone of the Van Allen radiation belt, *J. Geophys. Res.*, **65**, 3117–3130.

Piddington, J. H., and Drake, J. F., 1968: Electrodynamic effects of Jupiter's satellite Io, *Nature*, **217**, 935–937.

Richardson, J. D., Siscoe, G. L., Bagenal, F., and Sullivan, J. D., 1980: Time dependent plasma injection by Io, *Geophys. Res. Lett.*, **7**, 37–40.

Russell, C. T., Baker, D. N., and Slavin, J. A., 1988: The magnetosphere of Mercury, in F. Vilas, C. R. Chapman, and M. S. Matthews (eds.), *Mercury*, pp. 514–561, Univ. of Ariz. Press, Tucson.

Schulz, M., 1976: Plasma boundaries in space, in D. J. Williams (ed.), *Physics of Solar Planetary Environments*, vol. 1, pp. 491–504, Am. Geophys. Union, Washington, DC.

Schulz, M., 1991: The magnetosphere, in J. A. Jacobs (ed.), *Geomagnetism, vol. 4*, ch. 2, pp. 87–293, Academic Press, London, UK.

Schulz, M., and McNab, M. C., 1995: Source-surface modeling of planetary magnetospheres, *J. Geophys. Res.*, **100**, submitted for publication.

Schulz, M., Blake, J. B., Mazuk, S. M., Balogh, A., Dougherty, M. K., Forsyth, R. J., Keppler, E., Phillips, J. L., and Bame, S. J., 1993: Energetic particle, plasma and magnetic field signatures of a poloidal pulsation in Jupiter's magnetosphere, *Planet. Space Sci.*, **41**, 967–975.

Schulz, M., McNab, M. C., Lepping, R. P., and Voigt, G.-H., 1995: Magnetospheric configuration of Neptune, in D. P. Cruikshank (ed.), *Neptune and Triton*, in press, Univ. of Ariz. Press, Tucson.

Selesnick, R. S., 1988: Magnetospheric convection in the nondipolar magnetic field of Uranus, *J. Geophys. Res.*, **93**, 2607–2620.

Selesnick, R. S., 1990: Plasma convection in Neptune's magnetosphere, *Geophys. Res. Lett.*, **17**, 1681–1684.

Selesnick, R. S., and Richardson, J. D., 1986: Plasmasphere formation in arbitrarily oriented magnetospheres, *Geophys. Res. Lett.*, **13**, 624–627.

Simpson, J. A., Bastian, T. S., Chenette, D. L., McKibben, R. B., and Pyle, K. R., 1980: The trapped radiations of Saturn and their absorption by satellites and rings, *J. Geophys. Res.*, **85**, 5731–5762.

Singer, S. F., 1958: Trapped albedo theory of the radiation belt, *Phys. Rev. Lett.*, **1**, 181–184.

Siscoe, G. L., 1977: On the equatorial confinement and velocity space distribution of satellite ions in Jupiter's magnetosphere, *J. Geophys. Res.*, **82**, 1641–1645.

Siscoe, G. L., 1979: Towards a comparative theory of magnetospheres, in Solar System Plasma Physics, C. F. Kennel, L. J. Lanzerotti, and E. N. Parker (eds.), *vol. 2*, pp. 319–402.

Siscoe, G. L., and Summers, D., 1981: Centrifugally driven diffusion of Iogenic plasma, *J. Geophys. Res.*, **86**, 8471–8479.

Volland, H., 1975: Models of global electric fields within the magnetosphere, *Ann. Géophys.*, **31**, 171–180.

Whang, Y. C., and Ness, N. F., 1975: Modeling the magnetosphere of Mercury, *Rept. NASA-GSFC-X-690-75-89*.

Wu, C. S., and Lee, L. C., 1979: A theory of the terrestrial kilometric radiation, *Astrophys. J.*, **230**, 621–626.

*constituting a small but representative sample of literature on the subject reviewed.

BOUNDARY DETERMINATIONS FROM LOW FREQUENCY MAGNETIC FIELD MEASUREMENTS

L.J. ZANETTI, T.A. POTEMRA, B.J. ANDERSON
Johns Hopkins University, Applied Physics Laboratory
Johns Hopkins Road, Laurel, Maryland 20723-6099

Objects in the interplanetary medium are subject to interactions with the solar wind. As the solar wind encounters magnetized or unmagnetized objects, a shock wave forms upstream of the object and the solar wind field is distorted and draped over the obstacle. For magnetic bodies, this interaction produces three distinct regions of space: the solar wind, the magnetosphere of the body, and the magnetosheath or shocked solar wind between the solar wind and the magnetosphere. Accurate characterization of magnetic bodies therefore depends on identification of crossing from the solar wind or magnetosheath into the magnetosphere of the object. Magnetic measurements in the < 1Hz to 100's Hz range (denoted AC) have proven extremely useful in determining characteristic regions within the Earth's magnetosphere. At low altitudes, magnetic fluctuation levels are particularly useful in identifying field-aligned currents of the ionosphere auroral zones. These currents are a necessary consequence of the solar wind and will occur at other bodies as well, connecting the various altitude regions of magnetospheres. At high altitudes, the AC levels have proven indicative of magnetosheath and magnetopause current layers. Magnetosheath regions are subject to ion cyclotron and mirror mode local instabilities which do not penetrate the magnetopause. Furthermore, the magnetopause boundary for large magnetic shear is the site of intense magnetic noise. Magnetic fluctuation levels may be used to assist in the identification, at various altitudes, of the boundaries between the regions of magnetic influence of the solar wind and interplanetary bodies.

An exciting new discovery of the field-aligned currents is the AC frequency wave power or structure associated with the various large-scale current systems (*Anderson et al.*, 1993). A deflection of the component parallel to the main field is consistent with the associated auroral electrojet and other ionospheric Hall currents (*Zanetti et al.*, 1983). The Freja mission (600 × 1700 km, 63° inclination orbit, launched on October 6, 1992) is a Swedish scientific mission concentrating on auroral physics and involved instruments with which to measure fields and plasma. The Johns Hopkins University Applied Physics Laboratory (JHU/APL) provided the Magnetic Field Experiment.

Figure 1 is a polar plot in magnetic coordinates of the disturbance field transverse to the Earth's IGRF90 model magnetic field on April 4, 1993, interpreted as field-aligned currents. Time series disturbances in magnetic coordinates, north (BN), east (BE) and

Earth, Moon, and Planets **67**: 175–178, 1995.
© 1995 *Kluwer Academic Publishers.*

176

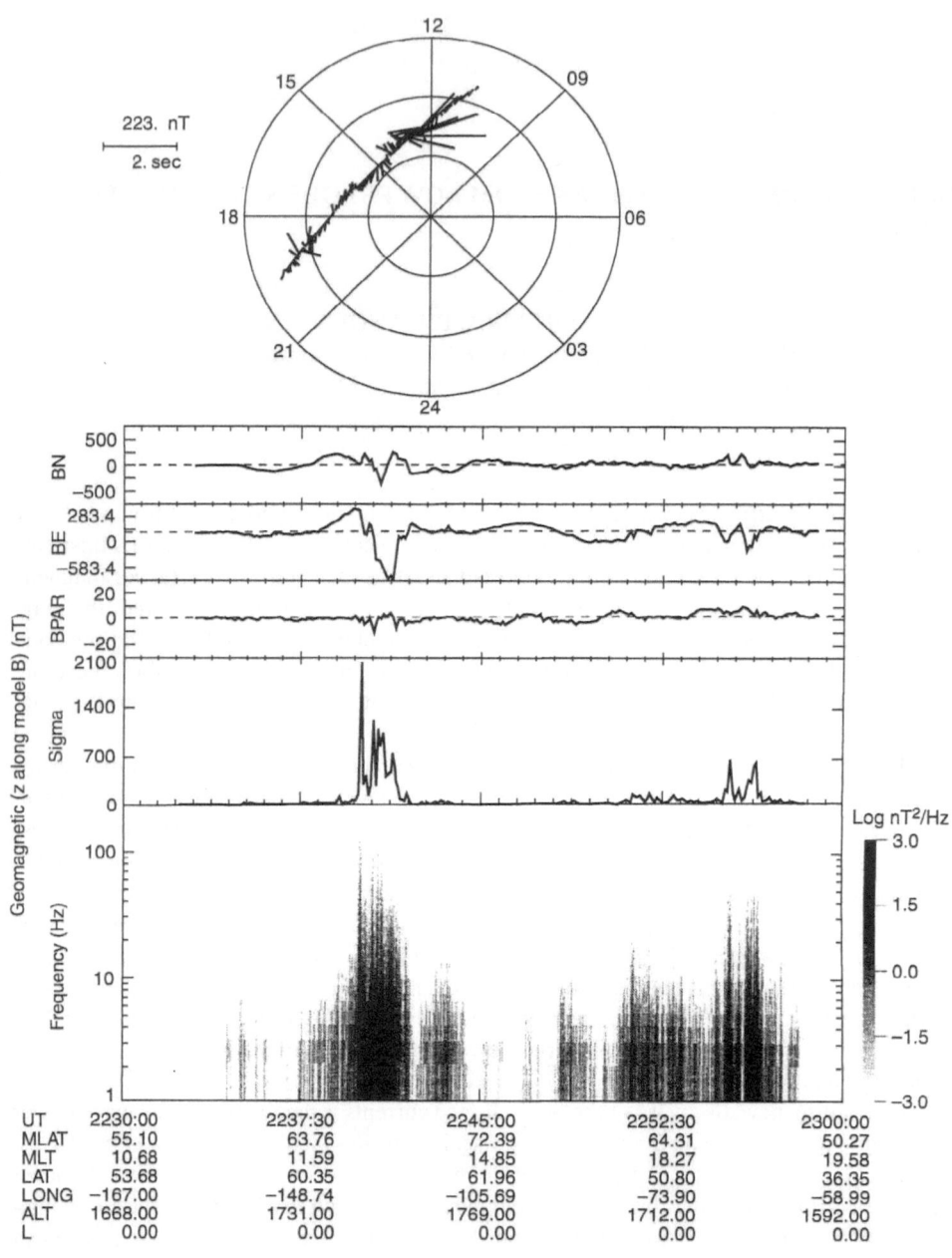

Figure 1. Freja Magnetic Field, April 4, 1993.

94-2875

parallel (BPAR) are also plotted. The next panel shows onboard processed standard deviation (nT) of the spin axis 1.5–128 Hz bandpassed channel showing the correlation to the auroral current zone. The bottom panel shows the onboard processed FFT in spectrogram format, 1–256 Hz, versus time; power (log nT^2/Hz) is indicated by greyscale. Details of the broadband wave activity correspond to the structure of the auroral zone currents. The Sigma and the FFT panel show the correlation of fluctuations, or characteristic scale lengths, of the current structure with the large-scale transverse vector disturbances which have traditionally been used to identify the field-aligned currents, which can more simply identify the auroral current zone.

The Active Magnetospheric Particle Tracer Explorers (AMPTE) program consists of three spacecraft (US, Germany, UK) and was launched in August 1984. The JHU/APL Charge Composition Explorer (CCE) was in an equatorial, 8.8 Re apogee, elliptical orbit and entered the magnetosheath and even the solar wind during large compressions of the magnetosphere.

Figure 2 (*Anderson and Fuselier*, 1993) shows AMTPE/CCE magnetic field data on December 13, 1984 from 02:00 UT to 05:00 UT during an exit of the Earth's magnetosphere into the magnetosheath. The magnitude and polar angles are shown in the left panel with the magnetopause (MP) identified. The FFT spectrograms in the right panel show transverse electromagnetic ion cyclotron and compressional mirror waves, both clearly

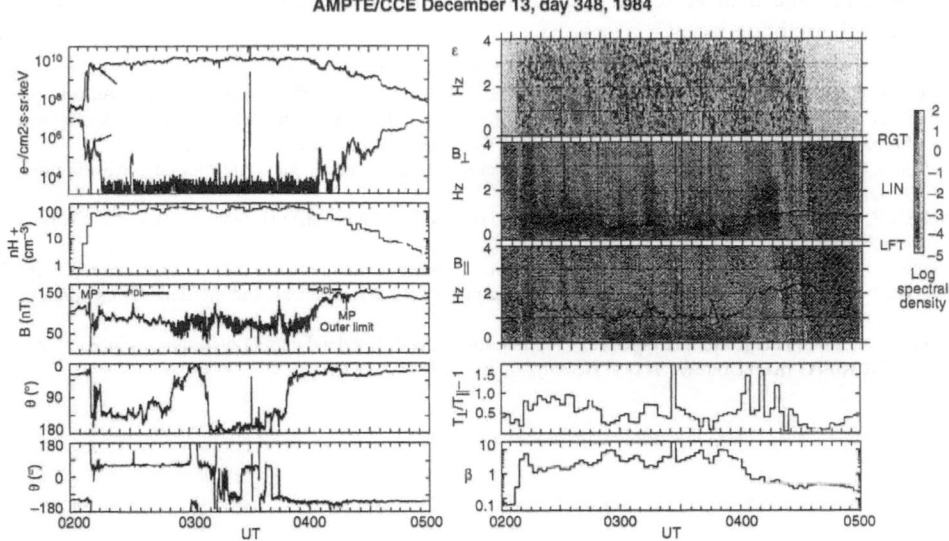

AMPTE/CCE December 13, day 348, 1984

Figure 2. AMPTE/CCE magnetic field data on December 13, 1984, from 0200 UT to 0500 UT during an exit of the Earth's magnetosphere into the magnetosheath. The magnitude and polar angles are shown in the left panel with the magnetopause (MP) identified. The FFT spectrograms in the right panel show transverse electromagnetic ion cyclotron and compressional mirror waves, both clearly identifying the magnetosheath area. The dark, low-power region indicates the quiet magnetosphere.

identifying the magnetosheath area. The dark, low power region indicates the quiet magnetosphere.

Studies of magnetic fluctuation levels have led to much increased understanding of the magnetosheath and magnetopause. The outbound magnetopause crossing occurred at 02:10 UT. The inbound crossing occurs for low magnetic shear but discontinuous changes in the 50–300 eV electrons indicate that the crossing was at 04:17 UT. The background fluctuation level is more intense in the magnetosheath (03:30–04:00 UT) than in the magnetosphere (04:40–05:00 UT). Transverse and compressional band limited signals are also apparent. Lastly, broadband perturbations occur during the high shear magnetopause crossing. As shown, the 50–300 eV electrons display discontinuous changes even during low-shear magnetopause crossings when the current layer is impossible to identify locally in non-AC magnetic field data. The final observation of the magnetopause region is the broadband noise signal which appears to be inherent to the magnetopause current layer. The outbound and inbound magnetopause crossings of Figure 2 illustrate the contrast in fluctuation levels between high and low magnetic shear. *Drake et al.* (1994) have found that current layers are susceptible to an instability in which the current layer degenerates into a turbulent structure of filamentary currents.

In summary, low frequency magnetic field fluctuations can be used to identify boundaries of magnetospheres and solar wind interaction regions due to high altitude magnetopause and the outer magnetosphere magnetic field structure. At low altitudes, high latitude AC structure is due to the current carrying packets, thus making the fluctuation characteristics more reliable than vector field changes to identify the current regions which define magnetospheric boundaries.

Anderson, B.J., Potemra, T.A., Bythrow, P.F., Zanetti, L.J., Holland, D.B., and Winningham, J.D., Auroral currents during the magnetic storm of November 8 and 9, 1991: observations from the Upper Atmosphere Research Satellite Particle Environment Monitor, *Geophys. Res. Lett.*, **20**, 1327, 1993.

Anderson, B.J., and Fuselier, S.A., Response of thermal ions to electromagnetic ion cyclotron waves, in press, *J. Geophys. Res.*, 1994.

Drake, J.F., Gerber, J., and Kleva, R.G., Turbulence and transport in the magnetopause current layer, *J. Geophys. Res.*, **99**, 11,211–11,223, 1994.

Zanetti, L.J., Anderson, B.J., Potemra, T.A., Kappenman, J., Lesher, R., and Feero, W., Ionospheric currents correlated with geomagnetic induced currents: Freja magnetic field measurements and the sunburst monitor system, *Geophys. Res. Letters*, **21**, 1867, 1994a.

Zanetti, L.J., Baumjohann, W., and Potemra, T.A., Ionospheric and Birkeland current distributions inferred from the MAGSAT magnetometer data, *J. Geophys. Res.*, **88**, 4875, 1983.

Evolution of Planetary Ringmoon Systems

Jeffrey N. Cuzzi
Space Science Division, Ames Research Center, NASA
Moffett Field, California 94035 USA

January 20, 1995

Abstract. The last few decades have seen an avalanche of observations of planetary ring systems, both from spacecraft and from Earth. Meanwhile, we have seen steady progress in our understanding of these systems as our intuition (and our computers) catch up with the myriad ways in which gravity, fluid and statistical mechanics, and electromagnetism can combine to shape the distribution of the submicron-to-several-meter size particles which comprise ring systems [1-5]. The now-complete reconnaissance of the gas giant planets by spacecraft has revealed that ring systems are invariably found in association with families of regular satellites, and there is an emerging perspective that they are not only physically but causally linked. There is also mounting evidence that many features or aspects of all planetary ring systems, if not the ring systems themselves, are considerably younger than the solar system.

Key words: Planetary Rings, Outer Planets, Origin and Evolution

1. Origin and Evolution

The fundamental goals of ring studies are to understand the origin of ring systems, and to use them as dynamical analogs of astrophysical particle disks in general. The origin of rings has challenged theorists for two centuries; in essence, explanations all relate to the effects of planetary tidal forces. However, consensus has shifted repeatedly over the years between the idea that rings come from moons which were torn asunder by the planet's gravity or by impact (dating from Roche), and the idea that rings are primordial remnants unable to accrete within the zone where tidal forces overwhelm the self gravity of growing satellites [6]. Current understanding favors the "destruction" model in which rings are derivative. In either case, to understand ring origin we must peer back through the evolutionary processes that have acted on the rings and their associated ringmoons to bring them to their current state. We seek evidence of the nature of these processes in the current structure of the rings. A basic property of rings is their "optical depth" τ, which measures the extinction of radiation by material in the rings (see [1-5]). Large optical depths may be regarded either as the approximate number of times a photon would encounter a particle while passing normally through the ring; small optical depths approximte the fractional area filled by particles, or the probability a photon would encounter a particle.

Earth, Moon, and Planets **67**: 179–208, 1995.

2. Important processes

Of course, rings are merely an ensemble of individual objects in orbit about their parent planet. Acting on this ensemble are the handful of processes which, in our current understanding, have the major influence on ring structure.

2.1. VISCOSITY

Collisions between ring particles occur on time scales from small fractions of an orbit to many years, depending on the local optical depth of the rings. The orbiting particles attain random *relative* velocities due to a combination of physical collisions with their (differentially orbiting) neighbors and gravitational scatterings by the largest members. These random velocities act in a statistical mechanical sense to provide a viscosity ν; in fact, much of ring structure has been studied in terms of the behavior of a viscous fluid. In principle, reliable estimates of ring viscosity could constrain the physical nature of individual particles (compact ice balls or fluffy, easily fragmented temporary agglomerations of debris) and the variation throughout the rings of the balance between forcing and damping processes. However, even the physical behavior of the viscosity is not yet fully understood. "Particle in a box" statistical mechanics is not completely valid in these systems, due to the coupling of the velocity of a particle (and thus its "random" relative velocity at the point of collision with a neighboring particle) and its position in its orbit. Furthermore, theoretical studies have suggested that, as the particle number density increases, the collective properties of the ring particles can resemble those of a liquid more than those of a gas, and ultimately even "solid" phases may "freeze out" at least in transient regions [7]. Some evidence for this may be found in discrepancies being seen in careful radiative transfer modeling of the rings. Their photometric properties in many cases deviate from those of a layer of low volume density, as if the particles in some regions are more closely packed than in others [8]. Only very recently are the many simplifying assumptions which have characterized these studies being relaxed [9], and realistic collisions, particle size distributions, and gravitational scatterings by the larger particles included. Nevertheless, detailed inferences as to particle properties, energy budgets, and ultimately timescales in the real rings from such a perspective remain elusive. Further background on this general subject may be found in [10].

2.2. GRAVITATIONAL FORCES

Long before the Voyager encounters, it was realized that the relatively tiny gravitational forces of both nearby and remote satellites, with fractional mass $\mu \sim 10^{-8}$ that of the planet (or even less), could lead to significant effects at resonance locations where the orbital frequencies of the satellite and the ring particles are commensurate (integer fractions or multiples) to a precision on the order of $\mu^{1/2}$ (the "width" of the resonance). Initial studies of individual resonances borrowed from galactic dynamics, and emphasized Lindblad resonances - those between the radial oscillation period of the particle and the period of the perturber. These were quickly shown to explain the numerous spiral density waves seen in the rings of Saturn [2,11]. Subsequent studies extended this framework into vertical or inclination resonances, which lead to spiral bending waves or a flapping of the ring sheet [2,12], and corotation (angular motion) resonances [13]. Combinations of Lindblad and corotation types have been explored in the application to eccentric rings such as those of Uranus [14], and may be involved with angular confinement of ring material into arcs and clumps [15]. The general importance of collective effects in the transfer of angular momentum between moons and rings, regardless of the specific form of the effects (viscosity, self gravity, etc.) has been discussed in several very readable articles [16]. Analyses of density and bending wave profiles have been used to infer the ring mass density and viscosity in a dozen or so specific regions [17,2]. Kinematic viscosities are seen to vary throughout the rings, with lower values (probably $\sim (0.1 - 1)\tau$ cm^{2}sec^{-1}) in the C ring, and larger values ($\sim (10 - 100)\tau$ cm^{2}sec^{-1} in the A and B rings. These values are not far from those inferred from theoretical models of density wave damping given the expected particle sizes. Finally, gravitational perturbations in the presence of viscosity are the essence of what has become known as the "shepherding" process by which moons confine ring material, to which we return below.

2.3. METEOROID BOMBARDMENT

Although those who study the surfaces of the airless planets and satellites have long accepted the importance of extrinsic bombardment as a significant geological process, the importance of the neverending cosmic hailstorm has only recently gained its due attention in the context of ring systems. Actually, this process is probably even more important for the evolution of ring structure and composition than in the better studied case of surface cratering, because of the vastly greater surface area to mass ratio for ring systems than for moons. This process will appear several times in discussions below; articles dealing with this general subject are found in [18].

2.4. ELECTROMAGNETIC FORCES

Icy or rocky particles may become charged in the magnetospheres of the giant planets, and then experience Lorentz ($\mathbf{V} \times \mathbf{B}$) forces since their (Keplerian) orbital frequencies are in general different from the rotational frequency of the planetary magnetic field (that of the planet's mantle or deep interior). Because only a very tiny fraction of the mass in any of the main ring systems is in particles sufficiently small to be affected by electromagnetic forces (sizes of a micron or smaller), and such microscopic particles are extremely short-lived, these effects act primarily to redistribute recently generated dust. Nevertheless, these perturbations on the basic state are important for understanding the structure of the Jovian ring halo and Saturn's E ring, at least [19], and, while small, are unceasing and may play a role in long term ring evolution. For instance, it has been pointed out [20] that very fine charged grains, probably no more than molecular clusters, are unstable in Saturn's rings at radii between 1.53 and 1.63 R_S, where R_S is Saturn's radius, depending on their velocity. Interestingly, the 1.63 R_S limit does correspond to an abrupt change in several ring properties - not only optical depth [20], but also photometric behavior, typical radial structural scale, and presence or absense of spokes [21]. Larger charged grains receive a positive or negative torque from the planet's magnetic field, which rotates at a different rate than the ring particles except at "synchronous orbit" in the B ring. Landing at radii different from their source, they convey their new angular momentum to their new host particle which leads to radial drifts, as discussed further below. Further background in this area may be found in [22].

3. Ring Structure

3.1. Rings and ringmoons

Jupiter's ring system first revealed its presence to Pioneer 11 as a depletion of magnetospheric protons, and Voyager images provided its unambiguous detection (figure 1; [23]). The most recent studies of the Jovian ring reveal it as a relatively flattened belt containing both macroscopic and microscopic particles, transforming into a three dimensional "torus" of primarily microscopic grains inwards of the main ring, and into an extended, flattened, "gossamer" ring of much lower particle density ranging outwards of the main ring to beyond the orbit of Jupiter's innermost classical satellite Amalthea [24]. The presence of the microscopic dust and the inference of the macroscopic material led to the proposal of an ongoing process whereby micrometeoroid bombardment of a population of objects between centimeters and kilometers in size generates the visible microscopic dust, which is then redistributed and removed by a variety of processes on a timescale of $10^2 - 10^3$ years [25]. Searches of Voyager images have resulted in the discovery of at least two small moons orbiting in and around the Jovian main ring; these have now been complemented by groundbased images [1, 26]. One imagines these to be merely the largest of a distribution of objects ranging down to subkilometer size. Even though the orbits of the visible moons are now well determined, the geometry of the ring images is insufficiently accurate to pinpoint the locations of the moons precisely relative to the ring boundaries.

The first suggestion that nearby small moons could significantly influence ring dynamics and structure (dubbed "shepherding" by a member of the Voyager press corps) was in response to the discovery in 1977 of the Uranian rings during a stellar occultation by the planet [27]. For several years, the stability of these narrow, yet

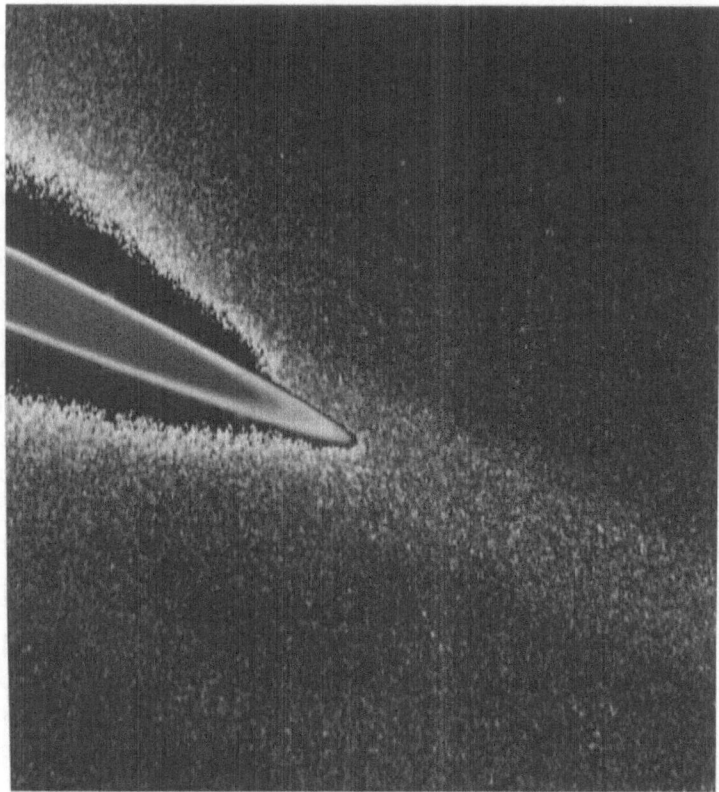

Fig. 1. A wide angle Voyager image, enhanced to display both the main ring and the outward extension, or "gossamer ring", which extends out well beyond the orbit of Jupiter's inner moon Amalthea. The vertical extension or "halo" material inwards of the main ring is also seen. The small moon Adrastea orbits close to the outer edge of the main ring, and its companion Metis lies within the ring. Figure from Showalter *et al.* (1985), in references [1].

quite dense and presumably collisionally active, rings remained a puzzle, because they were expected to spread radially and disperse due to their effective viscosity on a timescale $w^2/\nu \sim 10^5$ yr, where w is the ring width. The concept of shepherding [28, 29] is the transfer of energy and angular momentum between a ring and a more massive external moon (either nearby or remote) which counteracts the viscous spreading tendency. The essence of the Goldreich-Tremaine concept [28] has become generally accepted, although its specifics have evolved with time to keep pace with the observations.

The presense or prediction of "ringmoons" in and around the rings of Jupiter and Uranus was quickly echoed at Saturn during the Pioneer 11 and Voyager encounters of 1979 - 1981. Five substantial new moons were discovered skirting the edge of the main rings, and the classical satellite region is replete with debris. Lagrange point objects are seen in the orbits of Tethys and Dione [30]; in addition, unseen dispersed material is inferred in the orbits of these moons and those of Mimas, Enceladus, and Rhea as well as in regions devoid of any known moons of significant size [31].

The narrow, multistranded, kinky F ring lies between two of the five new inner moons; although this configuration has been referred to as the archetype of the shepherding process, it is not actually a particularly satisfying one. Most significantly, the ring may not be in torque balance; it is closer to the larger "shepherd" [32]. The lack of a good explanation for the presence of the F ring, and the presence of certain anomalous depletions of the inner magnetosphere surrounding it which do not correspond to any known rings or moons, has led Cuzzi and Burns [33] to suggest that the F ring is embedded within a much wider (1-2000km), but much more transparent (optical depth $\sim 10^{-3} - 10^{-4}$) ring or belt of asteroid-sized moonlets. This ensemble of objects is expected to interact collisionally to produce sporadic clumps of material; the F ring may be no more than an unusually large, rare collisional remnant. Recent analysis of azimuthal structure ("kinks" and "clumps") in the F ring appears to contain evidence for the presence of several members of the hypothesized moonlet belt population [34].

Inverting the original shepherding idea of Goldreich and Tremaine [28], it was suggested that the empty gaps in Saturn's rings were due to embedded small moonlets repelling ring material [35]. Initial attempts at direct detection of these objects were unsuccessful [36, 2], but indirect evidence for one such moonlet in the Encke gap of Saturn's A ring was accumulated [37], which allowed the object to be directly detected in fairly low resolution Voyager images [38]. Other gaps have been studied for indirect evidence of a similar nature, but the search has not been exhaustive and has not as yet met with comparable success [39].

The Voyager encounters with Uranus and Neptune completed the family portrait of the four ring systems (figures 2 and 3). The nine opaque rings of Uranus were found to be accompanied by about a hundred dusty bands of low optical depth (about 10^{-5}), and by ten new moonlets [40] of which one is embedded within the nine main rings. Similarly, Voyager found Neptune to have an extensive, low optical depth, ring system containing diverse elements with opacity ranging from 10^{-1} in the arcs, through 10^{-2} in two complete but narrow rings, to $10^{-3} - 10^{-4}$ in a broad, diffuse system about 30,000 km in width. Moonlets were also found in and around the Neptunian ring system with a radial distribution highly reminiscent of that found orbiting the other three gas giants. That is, they are found distributed throughout the observed ring material, both inside and outside of the Roche limit. Actually, the mass of the Neptunian ringmoons far exceeds that of the observed Neptunian rings, as seen in the Jovian ring-moon system. The orbits of two of the ring-related moonlets lie just about 1000 km inwards of the two major narrow Neptunian rings. This is probably not a coincidence, but as yet no one has grasped the significance of this configuration. It would be consistent with resonance trapping of inwardly drifting material (e.g., [41])

Fig. 2. **Voyager wide angle image of the Uranian ring system, obtained looking nearly directly** back towards the sun. This "forward scattering" geometry highlighted microscopic particles even of extemely low optical depths, and revealed for the first time that the nine known opaque, narrow rings were embedded in an extensive, structured, low (about 10^{-4}) optical depth system covering the entire region. The features in this image are about 50 km wide. Figure from Smith *et al.* (1986), in reference [97].

So, in the broadest perspective, we see a definite family similarity between the four ring-moon systems (figures 4 and 5). Each lies mostly within the Roche "zone" of its parent planet. The rings mingle with 1 - 10 embedded and outlying ringmoons, of tens to roughly a hundred km diameter, which themselves merge into the less numerous, larger "classical" moons further from the planet. The inner ring-moon systems are all prograde and equatorial - certainly not a foregone conclusion in the case of Neptune, which lacks a well-behaved classical satellite system.

Unfortunately, the Roche zone concept is actually not very well studied, as its applications lie in that messy regime so common in planetary science where realistic material properties strongly influence behavior and "spherical elephant" assumptions are glaringly inappropriate. The assumption of a liquid, or even self-gravitating, object is simply not adequate for irregular 10-100 km fragments, and

Fig. 3. Voyager wide angle, long exposure (about 10 minutes) image of the Neptunian ring system. This image is of fairly low resolution but shows the global structure of this optically thin $(10^{-4} - 10^{-2}$ except for the arcs which are $\sim 10^{-1})$ ring system. Several distinct components are visible, each having slightly different morphology and particle size distribution.

their actual properties, including internal strength in the presence of fractures and possible ice-rock boundaries, are not really known. Smoluchowski [42] pointed out that accretion of small particles onto the surface of larger ring particles of density ρ_{rp} becomes impossible at an "inner accretion limit" of $1.44(\rho_{pl}/\rho_{rp})^{1/3}R_{pl}$, where R_{pl} is the planet's radius and ρ_{pl} is its density; this is the location where (for characteristic particle spins) the combination of planetary tidal force, particle gravity, and centrifugal force just balance. This radius corresponds very well to the inner limits of all four ring systems (figure 4); one infers from this that particle histories in rings must result from a balance between accretion and erosion processes, since rings cannot long survive in the face of size-dependent removal processes where particles are no longer able to accrete, and thereby preserve, their smaller neighbors.

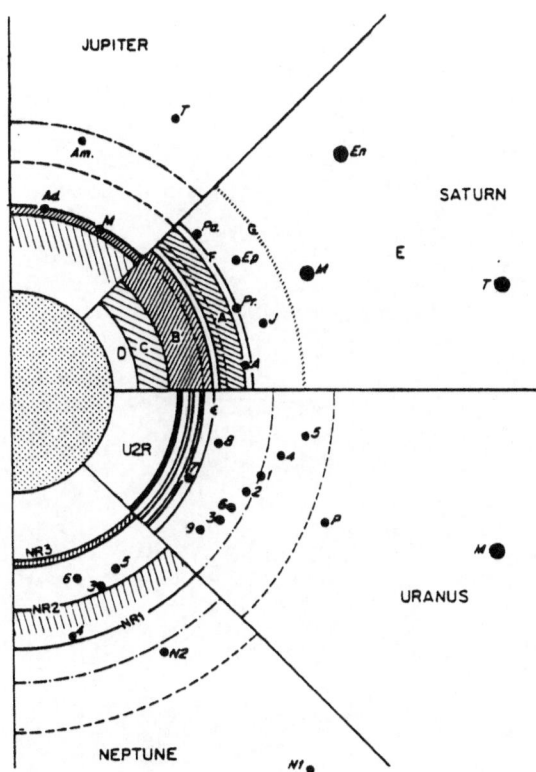

Fig. 4. A rendition of the intermingling of rings and ringmoons in the four systems, with radii scaled to the radius of the parent planet. The density of cross-hatching suggests the relative optical depths of the different rings. In each case, synchronous orbit is shown as a dashed line, and the Roche limit (for satellite density of 1 g cm^{-3} is shown by a dot-dashed line. (from Nicholson and Dones 1991, in reference [5].

Accretion of 10-100 km size objects from, say, a preexisting disk of small ring particles is unlikely anywhere in the rings due to the ability of such an object to repel surrounding material quite effectively. Objects this large must accrete outside the Roche limit. Once grown, reasonable material strength can allow moons of these sizes to survive within the outer Roche zones of their parents [43]. The implication is that moons form outside the Roche limit, and migrate into it. Occasional breakup of one of these objects can then provide ring material, and even entire rings, to replenish that which is continually lost. Source bodies could also be passing transients torn apart by planetary gravity [44]. A third possibility is that impact with preexisting ring material pulverizes an incident object, leading to capture of a larger amount of debris. One imagines that these processes recur throughout the ringmoon systems of the outer solar system.

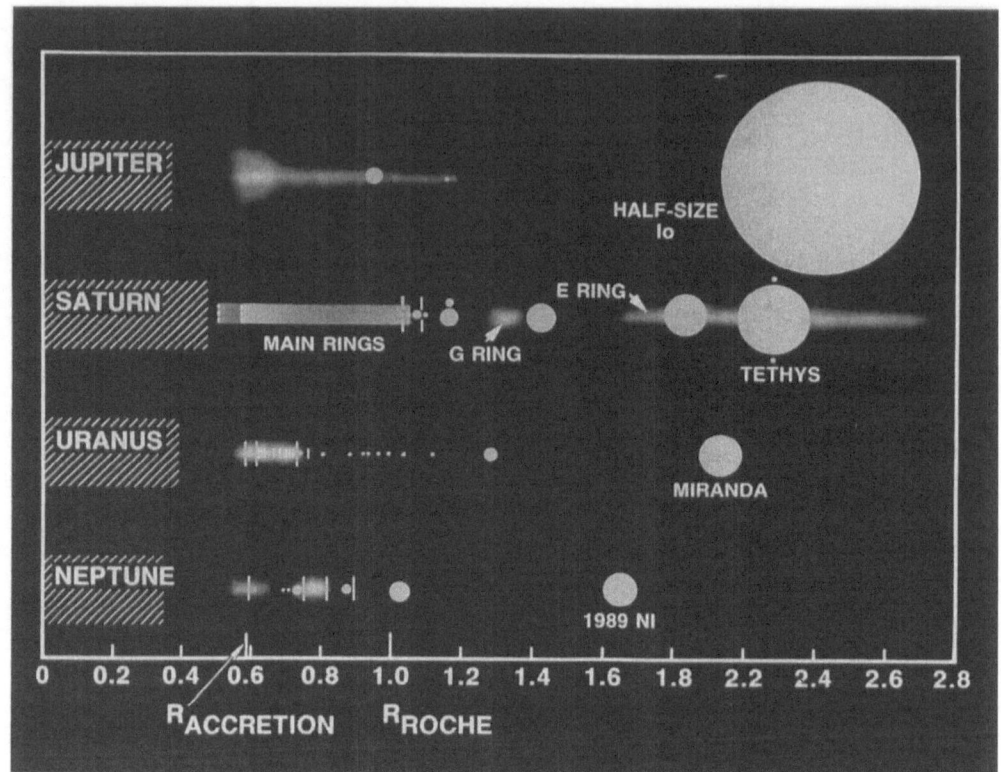

Fig. 5. A different rendition of the inner ringmoon systems of the four **gas giant planets, with** locations scaled by the "equivalent" radius $R_{eq} = (\rho_{rp}/\rho_{pl})^{\frac{1}{3}} R_{pl}$. In these units, the classical Roche limit lies at 2.44. The inner accretion limit is at 1.442 and is fairly closely related to the observed inner edges of all four ring systems (ref 38). The outer edges of most rings are in the vicinity of the classical Roche limit. Generally, the numbers of moons increase and their sizes decrease as the planet is approached in all cases.

3.2. SHEPHERDING

The ability of nearby moonlets to counteract the tendency of ring material to spread radially under viscosity has been a central theme of ring studies for nearly two decades now. The most cited early hypothesis [28] envisioned *local* satellites having numerous, overlapping resonances in the ring material. This concept, while containing the essential physics of angular momentum transfer, ignored the effects of the moonlet perturbations on the transport process. The torques involved are of the form

$$T = \nu \overline{\nabla v},$$

where ν is the viscosity and $\overline{\nabla v}$ is the velocity shear tensor, of the form $\left(\frac{\partial v_i}{\partial x_j} + \frac{\partial v_j}{\partial x_i}\right)$ where (i, j, k) denote radial, angular, and vertical coordinates. In early theories, the velocity shear was merely assumed to be Keplerian. That is, the interacting particles were assumed to lie on nearly circular orbits with velocity varying with radius due only to the planet's gravity. However, we now know [45] that the very perturbations caused by the moons alter the form of the velocity shear as a function of orbit longitude to the extent that at longitudes where the compression of material is the greatest, and the collisions are the most frequent, the *local* velocity gradient $\frac{\partial v_\theta}{\partial r}$ reverses from the Keplerian one and becomes positive over a restricted angular region. Overall, this causes the azimuthally averaged angular momentum transport and energy dissipation to be significantly decreased, and the ring material spreads much less rapidly in the presence of satellite perturbations than when unperturbed [45]. It is therefore much easier to "shepherd" ring material in this way than previously thought, especially narrow rings where the entire ring may lie in the perturbed region. The failure of Voyager to detect large embedded moons in some of the gaps in the rings of Saturn and Uranus is slightly less worrisome for this reason; however, there is fair agreement between the mass of the Encke moonlet Pan and the expected width of the Encke gap within current uncertainty in current parameter estimates. Consequently, unseen shepherds may yet be found in the Uranian system, but some concern remains about gaps in wide rings, such as at Saturn.

A slightly different "remote" shepherding process emerged when it was realized that two of the moonlets discovered by Voyager within the Uranian system have individual resonances at the edges of the largest Uranian (ϵ) ring [46]. The ability of isolated resonances to maintain ring edges had been previously discussed in the context of how Mimas delineates the outer edge of Saturn's B ring [47] and, in principle at least, in the context of the Uranian ring system itself [48]. Although no other moons have been found to provide confining torques for the other eight Uranian rings, exhaustive searches of the data to the level needed to approach the new lower masses required are difficult and time consuming, and remain to be done. There are locations from which several as-yet-undiscovered moonlets could influence the edges of several of the rings simultaneously; some of these are close to low-order resonances with known moons [49]. This implies that the unseen, locally controlling "shepherds" could be locked to larger objects.

The fact that partial "arcs" of ring-like material surrounded Neptune was first revealed by groundbased stellar occultations, and subsequently by Voyager images (figure 6, [50]).

Because of the differential orbital velocity across the width w of an arc at radius a from the planet, material with a spread w in orbital semimajor axes would ultimately spread to encircle the entire planet uniformly in a time roughly equal to a/w orbit periods - a matter of years. These arcs have been explained in terms of corotation resonance trapping by Galatea [51]. Jupiter's Trojan asteroid family is a

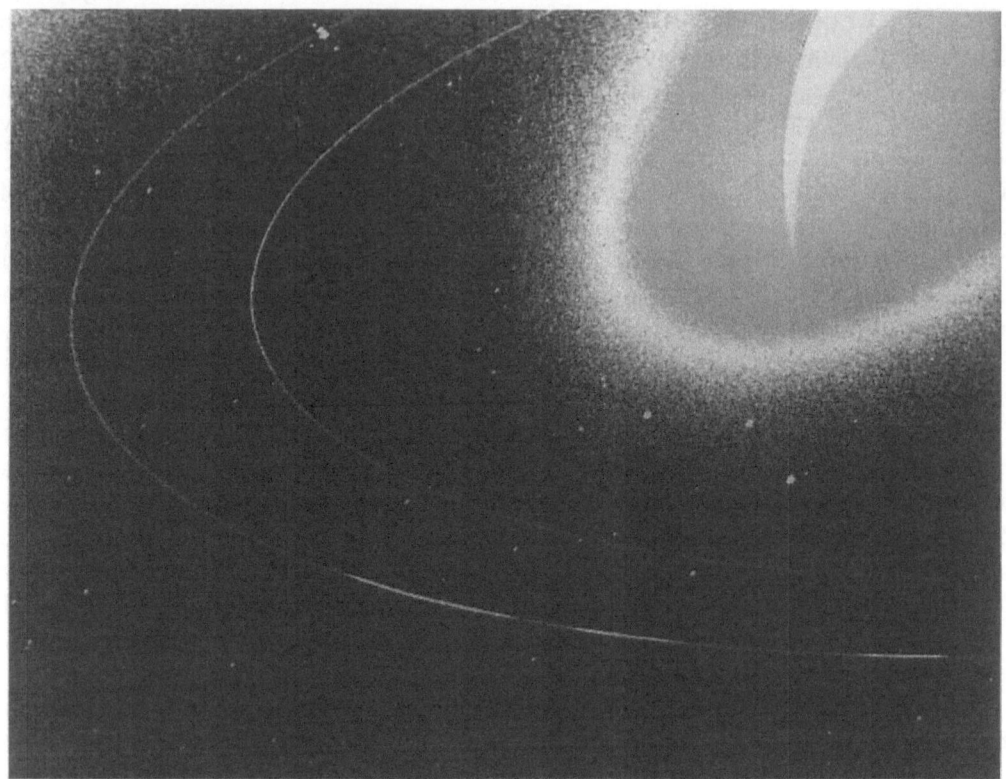

Fig. 6.   Wide angle image of the Neptune rings showing the three main arc features Liberté, Egalité, and Fraternité. These arcs can explain all of the groundbased detections of ring material around Neptune to date; one such observation actually detected a small moonlet (1989N1). Figure from Smith *et al.* (1989), in reference [97].

good example of a dissipationless system "confined" in the angular direction in the rotating frame of their prime perturber. The several arc confinement hypotheses which have been advanced thus far are variations on the theme of Lagrange point stability of (dissipationless) objects as coupled with some means of resupplying the energy lost by the colliding particles in the arcs [52]. Although the fundamental identification with the Galatea corotation resonances is probably secure, the Neptune arcs remain somewhat problematic - the observations are consistent with azimuthal confinement, but only if the spread in orbital semimajor axes is much smaller than the observed radial width of the arc material. It might be difficult to reconcile this situation with the expected level of collisional interactions, which excite random velocities [52]. Nevertheless, configurations have been proposed which minimize such spreading [53], and might be of relevance to the even smaller "clumps", with azimuthal scale of merely hundreds of km, which have also been found within the Neptune arcs [54]. An alternative possibility, dubbed "creationist" by one ring scientist, is that arcs in general might merely be transient, as suggested for the smaller clumps in Saturn's F ring region [33].

Other clumpy, azimuthally incomplete arcs are found in Saturn's F ring strands, the orbit of Neptune's ringmoon Galatea (1989N4), and also in Saturn's Encke gap. The Encke arcs are of somewhat larger angular extent than those of Neptune [2], and their longitudinal structure and relationship to the Encke moonlet have not been studied. Another new narrow, dusty, clumpy ring was discovered at Uranus, reminiscent of Saturn's F ring (λ; originally 1986U1R; [55]). Thus, while the main Neptune arcs are apparently confined in some way, others may be transient byproducts of interactions between local moonlets.

3.3. LOCAL STRUCTURE - MODES, WAVES, AND WAKES

Prior to Voyager encounter in 1986, several of the rings of Uranus were known to exhibit unexplained sub-km radius and/or width residuals from smooth, elliptical, inclined orbits. These residuals were identified [56] as patterns with low azimuthal wavenumber - an $m = 0$ or an axisymmetric "breathing" mode, and an $m = 2$ pattern such as excited along the outer edges of Saturn's B ring [47]. These patterns have a characteristic angular velocity for each m that, when included in the fits, reduce the residuals considerably. The possibility of free or excited modes in planetary rings was first suggested by Borderies and coworkers [57], who suggested that stable modes would be most likely to occur in rings where the material was extremely closely packed, with viscous properties more like those of a liquid than the gas-kinetic type viscosities more commonly adopted for sparser rings. The essence of the idea, as for favored modes in many other bounded systems (drumhead modes, for example) is that certain patterns of oscillation are stable due to the combination of their spatial and temporal scales of oscillation, and a balance between energy production and dissipation. Since the primary energy loss mechanism in rings is viscous coupling across the radial gradient in orbital velocity, normal modes are most stable in narrow rings, where this gradient can be most easily suppressed by the mode itself. The existence of elliptical modes which are stabilized by viscous forces might supercede the older idea that the elliptical shape of narrow rings is stabilized by the self-gravity of the ring; the self-gravity idea has led to some inconsistencies with observations [56]. In fact, all of the nine main rings of Uranus and several ringlets in Saturn's rings [2, 3, 39] show unexplained eccentricities ($m = 1$) which could be excited modes. Both the eccentricities and inclinations of the Uranian rings show a significant radial dependence [56, 58] which is not understood.

It is impossible to adequately address the full complexity of the structure in Saturn's ring system [2] in a brief article. In general, the three major ring regions (A, B, and C) exhibit different stucture (figure 7). The A (outermost) ring is characterized by a multitude of identified spiral density and bending waves, separated

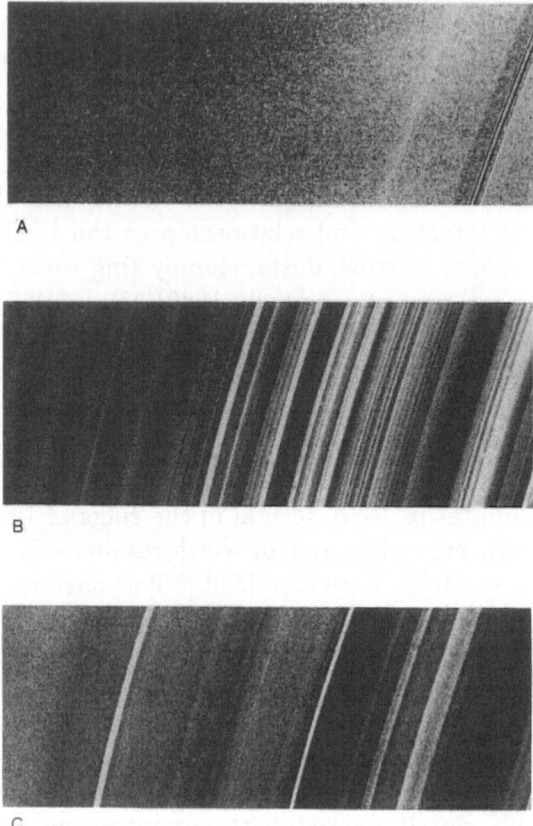

Fig. 7. Typical **structure in Saturn's rings varies between the classical ring** regions. The panels each display a region about 4000 km wide. The outermost (A) ring is generally featureless, but is punctuated with well separated and clearly identifiable trains of spiral density and bending waves excited at resonances with external moons. The middle panel shows the "irregular" structure filling the B ring. The lower panel displays the broad "plateau" stucture which characterizes not only the C ring shown, but also the Cassini division. The C ring and Cassini division are also the location of practically all of the empty gaps in the rings. Figure from Lissauer and Cuzzi (1985), in reference [5].

by regions which are featureless on short radial scales but exhibit a quadrupole azimuthal brightness asymmetry that has been ascribed to wake patterns which are on the scale of the ring thickness in horizontal scale [59]. One generally ascribes this behavior to the proximity of the A ring to the planet's Roche limit, where tidal effects weaken and coherent wakes are more easily generated. Salo [60] has shown that transient clumpings of particles occur on the tens-to-hundred meter scale even in the B ring, and long-lived agglomerations of several tens of meters radius can form in the A ring. The B ring, containing the bulk of the ring material, is filled with "irregular" radial structure on length scales of about 100 km. The characteristic length scale of this structure varies from ∼ 100 to ∼ 300 km; very

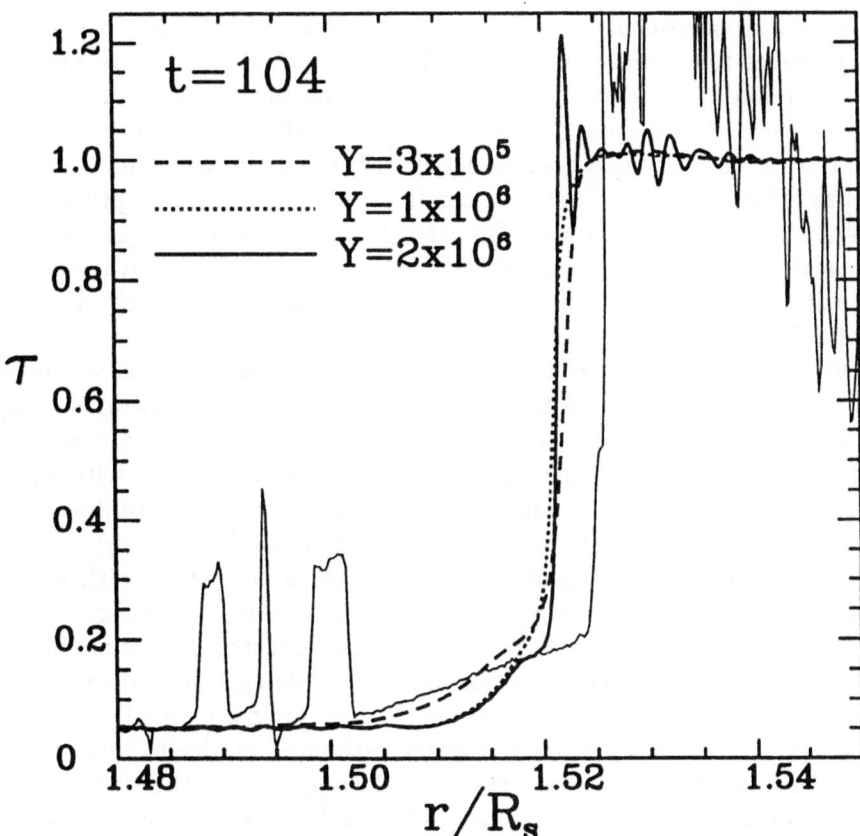

Fig. 8.   Structure generated at and near the inner edge of a step in optical depth similar in magnitude to that founud at the inner edge of Saturn's B ring, for three choices of a parameter Y that combines projectile flux and ring viscosity (heavy curves). The observed optical depth profile is shown by the lighter solid curve. The runs began with a smooth edge and cover a duration of 104 "gross erosion times". The gross erosion time is the time in which the incident flux generates a surface mass density of ejecta equal to that of the B ring (roughly 10^5 years for an ejecta yield of 10^5 times the projectile mass). Note that (a) ballistic transport maintains the edge in the face of the tendency of viscosity to make it spread, (b) a "ramp" forms inwards of the edge much like seen inwards of the observed A and B ring inner edges, and (c) "irregular" structure grows outwards of the edge, much as seen in the inner B and A rings. (from Durisen *et al.* 1992, in reference [65]).

fine scale structure ($\sim$ 10 km), some of which appears to be azimuthally variable, appears primarily in the outer 1000 km of the B ring [61, 2]. The lowest optical depth regions, the C (innermost) ring and Cassini division, are characterized by plateau-type structures with 100-1000 km radial scale, and empty gaps which often contain narrow ringlets. No explanation for the plateau-like structure of these regions is at hand.

Several hypotheses have been advanced to explain the B ring irregular structure. Some early ones invoked either very small moonlets, too small to completely clear gaps [35, 2], or viscous or collisional instabilities of various types, similar to those responsible for traffic jams [62, 2]. The viscous instability has encountered difficulties explaining the existence of characteristic length scales much larger than the ring thickness. Furthermore, it has now been realized that an important contribution to ring viscosity had been omitted from the early work: that due to finite particle size effects which are, in fact, dominant at large optical depth and probably prevent the instability [63]. On the other hand, others [64] suggest that stable eccentric features or "overstabilities" arising in large optical depth regions may produce azimuthal asymmetries, as in the case for isolated eccentric rings, on the scale of the finest structure

Two other explanations for the irregular structure have been advanced which are both related to transport of meteoroid ejecta. Durisen and coworkers [65] have shown that the optical depth dependence of the ability to release and absorb meteoroid ejecta may lead to growth of optical depth fluctuations with widths on the scale of the throw distance. The equilibrium amplitude of these fluctuations is given by a balance between ejecta yield and throw distance, and viscosity (which tends to diffuse the structure); observed structure is consistent with reasonable values of all parameters (figure 8). Goertz and coworkers [66] suggest a similar process, but relying on the electromagnetic torques incurred by grains temporarily charged in impact plasma clouds (discussed further below). Due to the more limited mass fraction of active carriers, this process creates structure only if viscosities are $\sim 10^{-2}$ (cgs). This is smaller than current estimates in the C ring, where the optical depth and particle size are both small compared to the rings in general; current estimates in the A ring are more in the 10-100 (cgs) range [17]; we have no observational viscosity constraints in the B ring, which is the location of nearly all of the irregular structure, but it is probably at least as large as in the A ring.

4. Ring particle properties

4.1. COMPOSITION

In light of the vast amount of spatial structure under study in ring systems, surprisingly little is known about ring composition. The Pioneer and Voyager cameras carried only broadband, visual wavelength filters, and had no capability for reflectance spectroscopy. Naturally, groundbased observations are of lower spatial resolution; however, they are still capable of providing many important compositional constraints.

Groundbased reflectance spectroscopy of Saturn's rings long ago demonstrated that the rings contain large amounts of water ice and a small admixture of (probably) silicate and (possibly) other non-icy material [2]. Groundbased radar and radio observations, even before Voyager encounter, required the particles to have a very high microwave reflectivity and a very low microwave emissivity [67]. Combining these properties with particle sizes allows *bulk* material refractive indices to be estimated. The particle size distributions inferred by these studies have been confirmed by Voyager [68, 69], and the non-icy component is thus constrained to make up less than about 10% of the rings [67], and perhaps no more than 1% [70] by mass. Other constraints on the makeup of *surficial* material have been obtained from analysis of Voyager photometry. In order for the macroscopic ring particles in the A and B rings to have a Bond albedo of about 0.5, the individual grains making up the surface must be nearly devoid of non-icy impurities, to a level of a few percent [71].

This is not easy to explain in light of the large amount of meteoritic material the rings must have absorbed (section 2) - primitive material consists of intimate mixtures of carbonaceous, silicaceous, and icy grains and is probably extremely dark, like the nucleus of comet Halley. The highly reflective classical satellites of Jupiter and Saturn presumably grew sufficiently large and hot during their formation to melt, allowing the lighter icy material to separate from denser, more refractory material, and form an icy outer shell of typically hundreds of km thickness. Early processing of Saturn's ring material through a large, differentiated satellite is consistent with the presence of compositional variations on both local and global scales in Saturn's rings, inferred from preliminary Voyager color and photometric data [2]. Catastrophic breakup of a differentiated moon would generate a compositionally inhomogeneous population of moonlet-sized fragments which might provide the source for the observed ring material. More systematic study of the observed variations is just beginning, and considerable analysis and modeling remains to be done. It does appear that local radial color variations are most prominent in the opaque regions of the B ring. Elsewhere, global color and albedo variations are fairly smooth and consistent with meteoroid infall and ballistic transport [72].

The Uranian ring particles are known to be extremely dark and essentially colorless, reminiscent of primitive materials such as comet nuclei or carbonaceous chondrites [73]. The reflectivities and colors of the nearby ringmoons are similar, but their reflectivities drop abruptly towards the planet. Laboratory experiments [74] suggest that, if the primordial composition of these objects included significant amounts of carbon-bearing ices, millenia of bombardment by trapped magnetospheric protons and electrons may have stripped the associated Hydrogen and/or Oxygen, leaving behind more complex, and highly absorbing, hydrocarbons. On the other hand, these objects may merely (or primarily) preserve the composition of primitive non-differentiated material, or be so thoroughly saturated with infalling meteoroidal material as to have lost their primordial composition. These latter explanations would not help explain the abrupt albedo drop with decreasing albedo, just where the magnetosphere might be expected to become important, in the case of Uranus.

In the case of the rings of Jupiter and Neptune, uncertainties as to the net amount of material present (optical depth) prevents determination of the absolute particle brightness; for Jupiter, relative spectral reflectivity measurements [2] show the ring material to be reddish, in contrast to the Uranian ring material. However, the entire inner Jovian system is polluted with reddish material ejected from the volcanoes of Io. In the case of the Neptune rings, color measurements are nonexistent. Preliminary photometric modeling [75] implies that the particles are quite dark, like those of the Uranian rings.

Taking the current results at face value, Saturn's rings appear unique in their high reflectivity as well as in their sheer abundance of material. The old question of "why is Saturn the only giant planet with rings" now becomes "why is Saturn the only giant planet with big, bright rings"? The new question is not much less puzzling than the old one. We return to this issue in the next section.

4.2. SIZE DISTRIBUTION

The physical nature of the individual particles is of interest primarily through its relationship to ongoing local dynamical processes. In general, the largest "typical" particles in rings are on the order of meters to tens of meters in size. The size of the largest typical particle varies dramatically with location [69]. Averaged over large radial regions in Saturn's rings, this largest size is typically seen as a cutoff in a powerlaw distribution of the form $n(r) \propto r^{-s}$, with s in the range between 2.5 and 3.5 [2, 68]. For these distributions, most of the mass of the rings lies near the upper size cutoff. However, the tiny fraction of mass residing in the microscopic dust is of great interest because dust is extremely short lived due to a variety of removal processes [1, 25] and is thus a diagnostic of relatively vigorous local dynamics.

The exception, Saturn's E ring, contains a quite narrow size distribution of micron-radius particles which was originally thought to be realted to unusual geological processes on the surface of the moon Enceladus [2, 19, 76] but now seems more likely to be a self-supported configuration [77]. The concept is that continual erosion of Enceladus is driven by a population of eccentric, charged, micron-radius grains which are highly selected for by the combined action of solar radiation pressure and electromagnetic forces.

The fraction of micron-sized dust in moderately opaque rings ($\tau > 0.1$) is quite low - in the main rings of Saturn and Uranus it is probably on the order of one percent or less in general [78]. This recent result, which was not anticipated in preliminary post-Voyager review articles [2] will require changes in several hypotheses of ring structure which adopted now outdated dust fraction estimates on the order of 10%. From the very limited photometry done so far of the low optical depth rings of Jupiter and Neptune, and of the Uranian dust bands, the fraction of dust appears to be higher: 50% or greater. Furthermore, Saturn's F ring [79] and Encke gap ringlets, the Uranian λ ring, and the Neptunian ring arcs are examples of rings of moderate optical depth (~ 0.1) which have an even larger dust fraction, indicative of vigorous ongoing creation.

Modeling which includes the sporadic creation of debris by collisions and meteoroid impacts, and the continual removal of debris by sweepup, demonstrates the diminished capacity of nearly transparent rings to sweep up the dust into the surfaces of their macroscopic ring particles compared to opaque rings. In the case of the Uranian and Neptunian dust bands and rings, ongoing collisions in belts of meter-to-kilometer-sized moonlets can generate the dust and sweep it up again sufficiently rapidly to confine it to the region of creation [80].

The macroscopic ring particles seem to be fairly sturdy, at least where their properties can be inferred. It has been suggested that the 5 to 10 meter "particles" that we observe with various techniques [68, 69] have no long term integrity but are merely passing clusters of smaller fragments, accumulating and collapsing on an orbital timescale. These objects, dubbed "DEBS", or dynamic ephemeral bodies, in one hypothesis [81], would have little or no strength or elasticity. Although ring particles might well be shattered and reaccreted on $1 - 10^4$ year timescales, difficulties arise with the DEB idea in its original form, which advocates extremely fragile objects with a strength several orders of magnitude lower than granular ice or snow in order that tidal forces alone may disrupt particles as small as 10 m in radius. However, photometric analysis in both unperturbed regions and nearby regions perturbed by spiral density waves has shown that, contrary to previous beliefs, optically thick rings are *not* dusty, either with or without relatively vigorous collisions, as one might expect if the particles are so easily and so often disintegrated and reassembled. Instead, Dones *et al.* (1993, in [78]) have shown that the particle surfaces are made smoother as they collide more vigorously, much like well-packed snowballs. Although the radiative transfer models being used to interpret ring brightness are still somewhat idealized and are currently being improved [8], the fractional optical depth of dust appears to be no more than about 1% on the average in any of the main rings of Saturn or Uranus [78]. In addition, the damping of most observed spiral density waves requires a viscosity typical of a particle random velocity of $10^{-2} - 10^{-1}$ cm sec^{-1}, which is comparable to the velocity shear across a 1 - 10 m radius ring particle, and can only be maintained if the corresponding particles are fairly elastic. A slight variation on the "DEBS" idea, with the destruction mechanism working entirely by collisions and not by tidal breakup, would appear to allow the particles to be of moderate strength while yielding sizes consistent with these observations [82]. As mentioned earlier, "DEB"-like objects of much larger size (several tens of meters in radius) are seen to form in numerical simulations which include a realistic hard-sphere size distribution up to 5 meters in radius. Some of these are really not even genuine aggregates but only temporary "wakes".

5. Short timescales

The first study of ring particles with short lifetimes came in relation to the Jovian ring. All of the (observed) particles tend to be microscopic and are removed or destroyed by various processes in 100 years or so, leading to a model characterized by ongoing creation (by meteoroid bombardment of a population of more massive,

long lived parent objects; [1]). Shortly after spiral density waves were identified in
the rings of Saturn, it was realized that the transfer of angular momentum between
the rings and the nearby satellites in these waves was so rapid that neither the
moons nor the rings could maintain their present positions for more than about 10^7
years [83]. A related dynamical argument involves Saturn's pair of moons known
as the "coorbital satellites", Janus and Epimetheus. These objects are in such close
orbits that they cause each other to shift back and forth across the orbital distance
of their relative center of mass. In the frame of the larger, the smaller executes a
"horseshoe" orbit of a particular angular amplitude. The amplitude of this angular
excursion, or libration, depends on the energy of the configuration, and it has been
shown that the energy will be damped by density wave interactions to yield a very
small amplitude for the libration, if the configuration is as old as the solar system
[84]. Most possible loopholes in these dynamical arguments have been closed by
improvements in our understanding of the angular momentum transfer process
[16]; however, there may be other possible ways out involving as-yet unknown
ways to transfer angular momentum and energy from the planet to the rings (eg.
[85]), or to resonantly lock the moons at or near their current locations.

An independent short timescale argument comes from the icy purity of Saturn's
rings, discussed earlier. The rings are bombarded by a constant infall of interplanetary debris, as is well documented for the Earth at the present time and manifested
on the cratered surfaces of airless planets and satellites for eons into the past. The
meteoroid population is quite primitive, abundantly endowed with carbonaceous
and silicate material. There is, of course, an uncertainty in the flux of this material at Saturn. The best flux estimate comes from experiments on the Pioneer 10
and 11 spacecraft, which suggest a nearly constant volume density out as far as
Saturn [86]. It is then possible to estimate that the mass flux of this material into
the rings over the age of the solar system as roughly equal to the current mass of
the rings, far more than permitted in light of the allowed impurity content of the
ring material. In addition to the flux uncertainty, one needs to worry about the
persistence of absorbing properties of the infalling material after impact at tens of
kilometers per second. If the non-icy material is all dissociated into atoms and lost
to the planet, or recombined into nonabsorbing forms (e.g. pure C, Si, Mg, etc.
oxides), then timescales may be lengthened somewhat. However, even assuming
only 10% of the infalling impurities retain their absorbing nature in the rings, a
ring age of only 10^8 years has been inferred by Doyle *et al.* [71].

The generally assumed value of meteoroid flux has some support from the depletion of electrons in the Saturnian ionosphere. This unusual situation has been
ascribed to influx of charge-scavenging water vapor molecules from the rings [87],
and the only mechanism capable of generating sufficient water vapor is meteoroid
impact [18]. This process implies a loss rate of water ice from the rings of about
10^{-15} g cm^{-2} sec^{-1}, comparable to the currently estimated meteoroid mass flux
into the rings (which amounts to about a ring mass in the age of the solar system). Impact produced vapor is usually estimated as comparable in mass to the

impactor, so this crude agreement is not unreasonable. The ionized fraction of this vapor is subject to immediate loss along magnetic field lines to the planet, and for mass fluxes about an order of magnitude larger than current estimates, a lifetime of $10^7 - 10^8$ years has been derived for the inner B ring due to this process [88].

If meteoroid bombardment does proceed at these rates, yet one more effect shortens the lifetime of material in all ring systems. Whereas relatively small particles ($r < 1$ cm) are easily destroyed by magnetospheric sputtering (at Jupiter), gas drag (at Uranus), or catastrophic impacts (generally), they can be replaced by new ejecta from larger objects. However, even meter-sized objects suffer from orbital decay due to absorption of the meteoroidal mass flux. If the infalling mass has an essentially isotropic orbital distribution (such as the well-known Oort cloud comets), as suggested by the Pioneer 10 and 11 results, it has no preferred direction and thus conveys zero net angular momentum to the rings. If its mass is primarily absorbed by the rings, as seems reasonable, the ensuing decrease in angular momentum per unit mass results in orbital decay. Particles with a centimeter-to-meter size distribution like those in the main rings of Saturn and Uranus drift inwards at a rate of about a centimeter per year [89]. There do not appear to be any moonlets in these regions capable of absorbing this amount of angular momentum [90]. Thus, the entire C ring of Saturn, and the inner rings of Uranus, would be expected to fall into their planets in about 10^8 years. This effect (which depends, of course, on the poorly known meteoroid flux) dominates the widely discussed gas drag torque on the Uranian rings for particles larger than about one centimeter, and applies equally to all four giant planets.

Additional evolutionary processes, involving electromagnetic loss mechanisms, also follow from the presence of microscopic particles (which are the only particles significantly affected by electromagnetic forces) in the rings. It is generally accepted that the radially extended "spokes" seen flickering across the face of the B ring result from temporary enhancements of the fractional abundance of microscopic dust particles from undetectable levels to about one percent [71], and that these enhancements derive from an electrostatic charging process possibly related to meteoroid impacts [91]. An alternative hypothesis for subsequent rapid radial motion of the charged plasma, responsible for levitating the grains from the regoliths of their parent particles over radial distances of 10^4 km in only a few minutes, has been suggested ($m=1$ electrodynamical instability; [92]).

Generally, the presence in, at least, the spokes, of charged 0.1 - 1 micron sized grains is fairly well accepted. These charged grains receive a net torque from the differentially rotating magnetic field during their lifetime, which is conveyed to the ring region in which they ultimately come to rest. This process is also said to lead to radial drifts on timescales much shorter than the age of the solar system, but suffers from uncertainties due to the poorly known spoke optical depth, dust particle size distribution, and equilibrium particle charge [93].

It might be said that there really are three independent arguments in the set above - dynamical ones relating to ring-moon torques, extrinsic ones related to meteoroid bombardment, and an electromagnetic one based on observed spoke properties. It is possible that, say, the meteoroid flux and the torque models are independently off by an order of magnitude or so in the right direction. However, many workers in the field now look at rings as systems which can not survive for longer than 10 -100 million years, and, unless we live in "the age of rings", must be created and recreated many times over the age of the solar system.

6. Ring origins

The weight of the current evidence does not *prove* the youth of any particular ring, much less that of all ring systems; however, it is highly suggestive and worthy of attention. Within the context of ongoing creation, a scenario for ring origin may be drawn that enjoys a certain plausibility in spite of its speculative nature. The general thread follows that of Harris ([94]).

The formation of the giant planets occurred at a time when particulate and gaseous phases coexisted in the nebula. It has been suggested several times in the past that the process of gas accumulation and collapse of the giant planets themselves may lead to the spinning off of a prograde, equatorial disk of (gaseous and embedded particulate) material (e.g. [95]). It has been claimed that accreting gaseous material prefers prograde circumplanetary orbits; on the other hand, incoming planetesimals show little preference for the prograde direction [96]. Universally prograde satellite-ring systems then imply that these objects probably accreted from an in-situ circumplanetary nebula. No solid material can survive the gas drag in this dense subnebula for long, unless it grows to a size of roughly 100 km or more. We still do not know the details of planetary growth between meters and moonlets, but it clearly has occurred. So, one presumes that near the end of the "nebular" phase of planetary formation, the giant planets were surrounded by a retinue of moons, possibly with a bias toward radii of at least 100 km, and, after a brief period during which the surrounding primordial gas/debris disks were dissipated by (as yet unknown) processes, little else. The smaller of these moons might be of uniform primordial composition, and the larger might be differentiated by the heat of their own formation. This ensemble of moons probably extended all the way in toward the planet due to ongoing radial evolution throughout the entire nebular stage.

The lifetime of a moon to catastrophic disruption is dependent on its size and on the bombarding flux. Moons larger than 100 km radius are especially resistant to breakup, because of their gravitational binding energy. Smaller moons are held together primarily by their material strength, and are destroyed by smaller meteoroidal projectiles which are considerably more numerous. However, this increasing flux is offset by the smaller target cross sections to a degree which depends on the projectile size distribution. The bombarding flux has varied greatly over the age of the solar system, and is smaller at the current epoch than during the period recorded on the surfaces of airless planets and satellites. From (admittedly uncertain) extrapolations of the number densities of large craters on the surfaces of the satellites of the giant planets, it has been shown [97, 98] that moons as large as 100 - 200 km radius receive impacts sufficiently large to destroy them at least several times during the period over which the observed cratering occurred (thought to be primarily in the first billion years of the solar system, based on age dating of lunar geologic features).

The fate of the fragments depends on the location of the parent bodies, and on the energy of the disruptive event. Far from the planet, tidal forces will not prevent the fragments from reaccumulating into a single (probably smaller) satellite. Thus, the five inner classical moons of Saturn may each have reaccumulated several times over in the early years of the solar system. In general, the energy of accumulation seems to have softened the moons sufficiently to allow them to relax into spherical shapes. Hyperion, an unusually irregular fragment, probably escaped this fate due to the gravitational dispersal of other fragments of its parent by Titan before reaccretion could occur [99]. Close to the planet, the interplanetary flux is focussed by the planet to a degree which depends on the relative velocity, given by the orbital distribution of outer solar system objects which contribute to the flux. Current estimates assume a mix of short- and long-period comets captured directly from the Oort cloud [97]. These have fairly large eccentricities and are not concentrated significantly. However, the presence of a large population of inwardly diffusing, low inclination and eccentricity, objects has been suggested [100] to account for the origin of (generally prograde) short period comets. This population, if present, is strongly gravitationally focussed by the planets because of its low relative velocity, and would overwhelm current estimates of the bombarding flux, especially within 2-3 R_{pl}. Some evidence against this population, at least as reflected in the small particle population which most strongly drives ring evolution, is provided by photochemical models of the water and carbon content of the stratospheres of Uranus and Neptune [101]. Consequently, it is difficult at this time to attach great confidence to lifetimes of moons of small and intermediate sizes in this region. However, taking our best current estimates at face value [97, 98], the average lifetime of a 10 km diameter moon at 2 planetary radii (at Neptune) is a few times 10^8 years, and the lifetime of a 100 m object is about ten times shorter. Lifetimes are about ten times longer at Saturn and three times shorter at Uranus, with Saturn's larger mass not quite compensating for the larger projectile flux at Uranus [97]. These lifetimes should be compared to the estimated lifetimes of $10^7 - 10^8$ years being currently estimated for the inner Uranian and Saturnian rings, as mentioned above.

The general case of the Saturn ring system is especially difficult to explain, as noted above, because of its large mass and compositional purity. The mass of Saturn's rings is about equal to the mass of Mimas. The expected lifetime against disruption for such a large object is close to the age of the solar system, making the creation of Saturn's rings in the last 10^8 years an improbable event [98]. A similar alternative (dating back to Roche), involving tidal disruption of a large "comet" suffering a close encounter with the planet, has recently been discussed in the light of current knowledge [102]. It appears to be about equally (un)likely, but doubles the probability that one or the other event actually occurred. Of course, the parent object would have had to have been significantly differentiated, and the bulk of the non-icy material would have to be preferentially removed or hidden from view, but these do not seem to be impossible difficulties. The fact that localized and regional scale compositional differences *are* seen in Saturn's rings is consistent with such a concept. In interesting recent developments, the densities of Saturn's inner ringmoons seem to be significantly less than that of solid water ice [103], suggesting that they may be merely rubble piles of loosely consolidated fragments of prior disruptions. Clearly, considerable effort needs to be devoted to these lines of evidence.

7. Comparative planetology

Practically all of the structural and compositional properties of planetary ring-moon systems can probably be explained by the *same* processes acting under slightly different boundary conditions; this makes them ideal for comparative study. Several successes of the past have been described above. In the future, several other aspects are worthy of study:

The question of transience *vs* confinement for any and all ring features should be approached objectively. Clumpy features which may help clarify this distinction are found in the systems of Saturn, Uranus, and Neptune, and few have been studied in detail. Comparing their structures and, in some cases, subsequent evolution, could tell us which are really transient or lead to new understanding about confinement.

If rings are transient, their creation and re-creation relies on impact disruption of small moons, and the rates at which this happens in the different systems are strongly tied to the projectile populations in the outer solar system. A better understanding of these objects (number densities and orbital configurations, from submillimeter to kilometer sizes) is greatly needed. This might be garnered from more extensive studies and interpretation of crater counts on the surfaces of the satellites. Stratospheric photochemistry might also provide constraints. Variation of this population between Jupiter and Neptune might provide new understanding to questions of the evolution of the Kuiper belt and the origin of short-period comets.

Improved theoretical modeling of viscous transport in particle disks, including realistic size distributions and all important processes, is required; a better understanding of viscosity may help us resolve many of these uncertainties regarding the B ring fine structure. Considerable information remains to be extracted from realistic radiative transfer modeling of regions in the rings of Uranus and Saturn;

it may be possible to infer their volume density, with important implications for local dynamics. Because different ring systems vary in particle size, optical depth, and state of excitation, a complete theory will only be developed by relating to all known ring systems.

Groundbased spectral reflectivity measurements of all planetary rings are needed to understand their origin and evolution. In the years to come, these should be feasible as far as Uranus. In the case of Saturn's rings, regional variations in important constituents can be mapped out. The stability and compositional purity of Saturn's rings for times longer than $10^7 - 10^8$ years remains an issue of special concern, due to the difficulty of avoiding contamination and the improbability of creating such a massive system so recently. Comparisons of these spectra with those of better studied objects will help trace the evolution of rings - are they dominated by infall? What is the nature of the reddish material in Saturn's rings and how does it differ from the reddish colorant in the Jovian rings?

In the long term, one expects the Cassini orbiter mission to remove most of the uncertainties currently besetting us in the area of inferring ring origin. Nevertheless, a considerable amount of preparatory work, both theoretical and observational, remains to be done in the next decade.

Acknowledgements

I owe a great debt to Dr. James B. Pollack, who was a stimulating and supportive mentor and colleague during the years in which I became deeply involved in studies of planetary rings. Although rings, of course, were only one of Jim's wide-ranging interests, he personally made several key creative contributions to the subject. His insight led to the realization that wavelength-sized ice particles could resolve the "radio-radar paradox" which had everyone else stumped 20 years ago. No doubt, Jim will be often and fondly remembered in the Cassini era for these and his many other contributions.

Bibliography

1. Jupiter: J. A. Burns, M. R. Showalter, and G. E. Morfill (1984), in "Planetary Rings"; Univ. of Arizona Press; J. A. Burns, M. R. Showalter, J. N. Cuzzi, and J. B. Pollack (1980), Icarus, 44, 339-360; M. R. Showalter, J. A. Burns, J. N. Cuzzi, and J. B. Pollack (1985); Nature, 316, 526-528; M. R. Showalter, J. A. Burns, J. N. Cuzzi, and J. B. Pollack (1987); Icarus, 69, 458-498

2. Saturn: J. N. Cuzzi, J. J. Lissauer, L. W. Esposito, J. B. Holberg, E. A. Marouf, G. L. Tyler, and A. Boischot (1984), in "Planetary Rings"; Univ. of Arizona Press; L. W. Esposito, J. N. Cuzzi, J. B. Holberg, E. A. Marouf, G. L. Tyler, and C. C. Porco (1984), in "Saturn"; Univ. of Arizona Press

3. Uranus: J. L. Elliot and P. D. Nicholson (1984) in "Planetary Rings"; Univ. of Arizona Press; R. C. French, P. D. Nicholson, C. C. Porco, and E. A. Marouf (1990), in

"Uranus"; Univ. of Arizona Press; L. W. Esposito, A. Brahic, J. A. Burns, and E. A. Marouf (1990), in "Uranus"; Univ. of Arizona Press. J. N. Cuzzi and L. W. Esposito, Scientific American, July 1987

4. Neptune: Porco, C. C., P. D. Nicholson, J. N. Cuzzi, J. J. Lissauer, and L. W. Esposito (1994) in "Neptune and Triton", University of Arizona Press, D. Cruikshank and M. S. Matthews, eds.; cf also J. Geophys. Res. 96, Supplement, October 30 1991; E. Stone and E. Miner (1989), Science, 246, 1417-1421; B. A. Smith et al. ibid., 1422-1450; A. L. Lane et al. ibid., 1450-1454

5. General review articles: P. Goldreich and S. Tremaine (1982), Ann. Revs. Astron. Astrophys., 20, 249-283; J. J. Lissauer and J. N. Cuzzi (1985), in "Protostars and Planets II", University of Arizona press; P. D. Nicholson and L. Dones (1991), Revs. Geophys., Supplement, U. S. National Report to the I.U.G.G., April 1991, 313-327; L. W. Esposito (1993) Ann. Revs. Earth Planet. Science; 21, 487-523; C. C. Porco (1995) Revs. Geophys., Supplement, U. S. National Report to the I.U.G.G., in press

6. Pollack, J. B., A. L. Summers, and B. Baldwin (1973), Icarus, 20, 263-278; J. B. Pollack (1975), Space Sci. Revs., 18, 3-93; J. B. Pollack and J. N. Cuzzi, Scientific American, November 1981

7. Araki, S. and S. Tremaine (1986), Icarus, 65, 83-109; J. Wisdom and S. Tremaine (1988), Astron. J. 95, 925-940; N. Borderies, P. Goldreich, and S. Tremaine (1985), Icarus, 63, 406-420; F. H. Shu, L. Dones, J. J. Lissauer, C. Yuan, and J. N. Cuzzi (1985), Astrophys. J., 299, 542-573

8. Dones, L., M. R. Showalter, and J. N. Cuzzi (1989), in "Astrophysical Disks, J. Sellwood, ed., p. 25-26; Cambridge. Univ. Press

9. Salo, H. (1991) Icarus, 90, 254-270; (1992a) Icarus, 96, 85-106; (1992b) Nature, 359, 619-621; Richardson, D. C. (1994) Mon. Not. Roy. Astron. Soc., in press

10. Stewart, G. R., D. N. C. Lin, and P. Bodenheimer (1984), in "Planetary Rings", University of Arizona press; R. Greenberg (1988), Icarus, 75, 527-539; cf. also G. R. Stewart and G. Wetherill (1988), Icarus, 74, 542-553

11. Franklin, F. W., M. Lecar, and W. Wiesel (1984), in "Planetary Rings", University of Arizona press; F. H. Shu (1984), ibid.

12. Shu, F. H., J. N. Cuzzi, and J. J. Lissauer (1983), Icarus, 53, 185; S. K. Chakrabarti (1989), Mon. Not. Roy. Ast. Soc. 238, 1381-1394

13. Goldreich, P. and S. Tremaine (1980) Astrophys. J., 241, 425-441; Sicardy, B. (1991), Icarus, 89, 197-219

14. Borderies, N., P. Goldreich, and S. Tremaine (1983), Astron. J., 88, 1560 - 1568; S. F. Dermott (1984), in "Planetary Rings", University of Arizona press; P. Goldreich and C. C. Porco (1987), Astron. J., 93, 730-737

15. Lissauer, J. J. (1985), Nature, 318, 544-545; P. Goldreich, S. Tremaine, and N. Borderies (1986), Astron. J. 92, 490; D. N. C. Lin, J. Papaloizou, and P. Bodenheimer (1987); M. N. R. A. S. 227, 75; Porco, C. C. (1991) Science, 253, 995-1001

16. Meyer-Vernet, N. and B. Sicardy (1987), Icarus, 69, 157-175; R. Greenberg (1983), Icarus, 53, 207-218;

17. Lissauer, J. J., F. H. Shu, and J. N. Cuzzi (1984), in "Anneaux des Planetes", Cepadues Editions, Toulouse; L. W. Esposito (1986), Icarus, 67, 385-392; S. K. Chakrabarti (1989), Mon. Not. Roy. Ast. Soc. 238, 1381-1394; P. A. Rosen, G. L. Tyler, E. A. Marouf, and J. J. Lissauer (1991), Icarus, 93, 25-44

18. Morfill, G. E., H. Fechtig, E. Grün, and C. K. Goertz (1983), Icarus, 55, 439-447; W.-H. Ip (1983), Icarus, 60, 547-552; R. H. Durisen (1984), in "Planetary Rings", University of Arizona Press; E. Grün, H. A. Zook, H. Fechtig, and R. H. Giese (1985) Icarus, 62, 244-272; J. N. Cuzzi and R. H. Durisen (1990), Icarus, 84, 467-501

19. Burns, J. A., L. E. Schaffer, R. J. Greenberg, and M. R. Showalter (1985), Nature, 316, 115-119; J. A. Burns and L. E. Schaffer (1989), Nature, 337, 340-343; see also refs 77.

20. Northrop, T. G., and J. R. Hill (1982), J. Geophys. Res. 87, 6045; W.-H. Ip (1983), J. Geophys. Res. 88, 819-822

21. Smith, B. A., et al. (1982); Science, 215, 504-537; Doyle, L. R., L. Dones, and J. N. Cuzzi (1989), Icarus, 80, 104-135; L. A. Horn and J. N. Cuzzi (1994), Icarus, submitted; Estrada, P. and J. N. Cuzzi, Icarus, in press, 1994.

22. Grün, E., G. E. Morfill, and D. A. Mendis (1984), in "Planetary Rings", University of Arizona Press; C. K. Goertz (1989), Revs. Geophys. 27, 271-292

23. Acuna, M. H., and N. F. Ness (1976), J. Geophys. Res., 81, 2917-2922; T. Owen, G. E. Danielson, A. F. Cook, C. Hansen, V. L. Hall, and T. C. Duxbury (1979), Nature, 281, 442-446; B. A. Smith et al. (1979), Science, 204, 951-971; B. A. Smith et al. (1979), Science, 206, 927-950

24. Showalter, M. R., J. A. Burns, J. N. Cuzzi, and J. B. Pollack (1985); Nature, 316, 526-528; Showalter, M. R., J. A. Burns, J. N. Cuzzi, and J. B. Pollack (1987); Icarus, 69, 458-498

25. J. A. Burns, M. R. Showalter, J. N. Cuzzi, and J. B. Pollack (1980), Icarus, 44, 339-360; Burns, J. A., M. R. Showalter, and G. E. Morfill (1984), in "Planetary Rings"; Univ. of Arizona Press

26. Nicholson, P. D. and K. Matthews (1991) Icarus, 93, 331-346

27. Elliot, J. L., E. W. Dunham, and D. J. Mink (1977), Nature, 267, 328-330; Millis, R. L., L. H. Wasserman, and P. Birch (1977), ibid., 330-331

28. Goldreich, P. and S. Tremaine (1979), Nature, 277, 97-99; N. Borderies, P. Goldreich, and S. Tremaine (1984), in "Planetary Rings", University of Arizona press.

29. see, e.g Dermott, S. F., C. D. Murray, and A. T. Sinclair (1979), Astron. J., 84, 1225-1234; Dermott, S. F. (1984) in "Planetary Rings", G. Greenberg and A. Brahic, eds, Univ. of Arizona Press, for a different viewpoint

30. Gehrels, T. et al. (1980), Science, 207, 434-439; B. A. Smith et al. (1981), Science, 212, 163-191; B. A. Smith et al. (1982), Science, 215, 504-537; J. J. Lissauer and J. N. Cuzzi (1985), in "Protostars and Planets II", University of Arizona press

31. Simpson, J. A. et al. (1980), J. Geophys. Res. 85, 5731-5762; J. A. Van Allen (1982), Icarus, 51, 509-527; J. A. Van Allen (1984), in "Saturn", University of Arizona Press

32. Showalter, M. R., and J. A. Burns (1982), Icarus, 52, 526-544; cf., however, J. J. Lissauer and S. Peale (1986), Icarus, 67, 358-374

33. Cuzzi, J. N., and J. A. Burns (1988), Icarus, 74, 284-324

34. Kolvoord, R. A., J. A. Burns, and M. R. Showalter (1990) Nature, 345, 695-697

35. Lissauer, J. J., F. H. Shu, and J. N. Cuzzi (1981) Nature, 292, 707-711; M. Henon (1981) Nature 293, 33-35; also (1984) in "Anneaux des Planetes", Cepadues-Editions, Toulouse, 365-384

36. Smith, B. A. et al. (1982); Science, 215, 504-537

37. Cuzzi, J. N. and J. D. Scargle (1985); Astrophys. J. 292, 276-290; M. R. Showalter, J. N. Cuzzi, E. A. Marouf, and L. W. Esposito (1986) Icarus, 66, 297-323

38. Showalter, M. R. (1991) Science, 351, 709-713

39. Flynn, B. C. and J. N. Cuzzi (1989) Icarus 82, 180-199; E. A. Marouf and G. L. Tyler (1986) Nature, 323, 31-35

40. Owen, W. M. and S. L. Synnott (1987) Astron. J. 93, 1268-1271

41. Weidenschilling, S. J., and D. R. Davis (1985) Icarus, 62, 16-29; Weidenschilling, S. J., and A. A. Jackson (1993) Icarus, 104, 244-254

42. Smoluchowski, R. (1979) Nature 280, 377-378

43. Dobrovolskis, A. R. (1990) Icarus, 88, 24-38

44. Dones, L. (1991) Icarus, 92, 194-203; Boss, A. P., A. G. W. Cameron, and W. Benz (1991) Icarus, 92, 165-178; Sridhar, S. and S. Tremaine (1992) Icarus, 95, 86-99; Asphaug, E. and W. Benz (1994) Nature, in press.

45. Borderies, N., P. Goldreich, and S. Tremaine (1983), Astron. J. 88, 1560-1568; also (1983) Icarus, 55, 124-132; also (1984), in "Planetary Rings", University of Arizona Press; also (1985) Icarus, 63, 406-420

46. Porco, C. C. and P. Goldreich (1987), Astron. J. 93, 724-729; Goldreich, P. and C. C. Porco (1987), ibid, 730-736

47. Borderies, N., P. Goldreich, and S. Tremaine (1982) Nature, 299, 209-211

48. Fridman, A. M. and N. N. Gor'kavyi (1989) Sov. Sci. Rev. A. Phys., 12, 289-346

49. Marouf, E. A. et al. (1988), B. A. A. S. 20, 845; Gresh, D. L. (1990), Thesis, Stanford University; Murray, C. and R. P. and Thompson (1990) Nature, 348, 499-502

50. Reitsema, H. J. et al. (1982), Science, 215, 289; C. E. Couvault et al. (1986) Icarus, 67, 126; W. B. Hubbard et al. (1986) Nature 319, 636; Smith, B. A. et al. (1989), Science, 246, 1422-1449

51. Porco, C. C. (1991) Science, 253, 995-1001

52. Porco, C. C., P. D. Nicholson, J. N. Cuzzi, J. J. Lissauer, and L. W. Esposito (1994) in "Neptune and Triton", University of Arizona Press, D. Cruikshank and M. S. Matthews, eds.; Sicardy, B., and J. J. Lissauer (1992) Adv. Sp. Res. 12, 81-95

53. Gor'kavyj, N. N. (1991) Sov. Astron. Lett 17, 428-432; Gor'kavyj, N. N. and T. A Taidakova (1991) Sov. Astron. Lett 17, 462-645; Gor'kavyj, N. N. T. A Taidakova, and N. M. Gaftonyuk (1991) Sov. Astron. Lett 17, 457-461

54. Smith, B. A. et al. (1989), Science, 246, 1422-1449; Ferrari, C. and A. Brahic (1994) Icarus, in press

55. Ockert, M. E., J. N. Cuzzi, C. C. Porco, and T. V. Johnson (1987) J. Geophys. Res. 92, 14969-14978; also (1988), B. A. A. S. 20, 854; J. E. Colwell et al. (1990) Icarus, 83, 102-125; Showalter, M. R. Icarus, submitted (1994)

56. French, R. C., J. A. Kangas, and J. L. Elliot (1986) Science, 231, 480-483; French, R. C., P. D. Nicholson, C. C. Porco, and E. A. Marouf (1990), in "Uranus"; Univ. of Arizona Press

57. Borderies, N., P. Goldereich, and S. Tremaine (1983), Astron. J. 88, 1560-1568; also (1983), Icarus, 55, 124-132; also (1985) Icarus, 63, 406-420

58. French, R. C., J. L. Elliott, and D. A. Allen (1982), Nature 298, 827-829

59. Colombo, G. P. Goldreich, and A. W. Harris (1976) Nature 264, 344-346; N. N. Gor'kavyi and T. A. Taydakova (1989), Pic'ma V. AZh, 15, 547-553 (in Russian, translation available; NASA TT-20576); L. Dones and C. C. Porco (1989) B. A. A. S., 21, 929

60. Salo, H, (1992) Nature, 359, 619-621

61. Horn, L. A., J. Hui, and J. N. Cuzzi (1989) B. A. A. S. 21, 928; L. A. Horn and J. N. Cuzzi (1994), Icarus, submitted

62. Lin, D. N. C. and P. Bodenheimer (1981), Astrophys. J., 248, L83; W. R. Ward (1981) Geophys. Res. Lett. 8, 641

63. Wisdom, J. and S. Tremaine (1988), Astron. J. 95, 925-940; Richardson, D. C. (1994) Mon. Not. Roy. Astron. Soc., in press; Borderies, N., P. Goldreich, and S. Tremaine (1985) Icarus, 63, 406-420

64. Borderies, N., P. Goldreich, and S. Tremaine (1983), Astron. J. 88, 1560-1568; also (1985) Icarus, 63, 406-420

65. Durisen, R. H., N. L. Cramer, B. W. Murphy, J. N. Cuzzi, T. L. Mullikin, and S. E. Cederbloom (1989) Icarus, 80, 136-166; Durisen, R. H., P. W. Bode, J. N. Cuzzi, S. E. Cederbloom, and B. W. Murphy (1992) Icarus, 100, 364-393

66. Goertz, C. K. and G. E. Morfill (1988), Icarus 74, 325-330; Goertz, C. K. (1990) Personal communication

67. Pollack, J. B., A. L. Summers, and B. Baldwin (1973) Icarus, 20, 263-278; J. N. Cuzzi and J. B. Pollack (1978) Icarus, 33, 233-263; J. N. Cuzzi, J. B. Pollack, and A. L. Summers (1980) Icarus, 44, 683-705; F. P. Schloerb, D. O. Muhleman, and G. L. Berge (1980) Icarus 42, 125-135; E. E. Epstein, M. A. Janssen, and J. N. Cuzzi (1984) Icarus, 58, 403-411

68. Marouf, E. A., G. L. Tyler, H. A. Zebker, and V. R. Eshleman (1983) Icarus, 54, 189 - 211; Zebker, H. A., E. A. Marouf, and G. L. Tyler (1985) Icarus, 64, 531-548

69. Showalter, M. R. and P. D. Nicholson (1990) Icarus, 87, 285-306

70. A. L. Grossman, D. O. Muhleman, and G. L. Berge (1989), Science, 245, 1211 - 1215.

71. Doyle, L. R., L. Dones, and J. N. Cuzzi (1989), 80, 104-135; L. Dones, personal communication (1990) (cf. also A. Grossman, Ph. D. Thesis, Caltech, 1988; A. Verbischer, P. Helfenstein, and J. Veverka (1990), Nature, 347, 162-164)

72. Estrada, P. and J. N. Cuzzi (1994), Icarus, in press; Cuzzi, J. N. and P. Estrada (1995) in preparation

73. Cuzzi, J. N. (1985) Icarus, 63, 212-215; M. E. Ockert, J. N. Cuzzi, C. C. Porco, and T. V. Johnson (1987) J. Geophys. Res. 92, 14969-14978; C. C. Porco, J. N. Cuzzi, M. E. Ockert, and R. J. Terrile (1987) Icarus, 72, 69-78; T. Svitek and G. E. Danielson (1987), J. Geophys. Res. ,92, 14979-14987

74. Strazzulla, G., L. Calcagno, and G. Foti (1983), M. N. R. A. S., 204, 59; L. J. Lanzerotti, W. L. Brown, and R. E. Johnson (1985), in "Ices in the Solar System", Reidel, Dordrecht; W. R. Thompson, B. Murray, B. N. Khare, and C. Sagan (1987), J. Geophys. Res. 92, 14933

75. Ferrari, C. and A. Brahic (1994), Icarus, in press; Porco, C. C., P. D. Nicholson, J. N. Cuzzi, J. J. Lissauer, and L. W. Esposito (1994), in "Neptune and Triton", University of Arizona Press.

76. Pang, K. D., C. Voge, J. W. Rhoads, and J. M. Ajello (1984) J. Geophys. Res. 89, 9459-9470; Showalter, M. R., J. N. Cuzzi, and S. M. Larson (1991), Icarus, 94, 451-473

77. Horanyi, M., J. A. Burns, and D. Hamilton (1992) Icarus, 97, 248-259; Hamilton, D. and J. A. Burns (1994); Science 264, 550-553.

78. Ockert, M. E., J. N. Cuzzi, C. C. Porco, and T. V. Johnson (1987) J. Geophys. Res. 92, 14969-14978; Dones, L., Cuzzi, J. N., and M. R. Showalter (1993) Icarus, 105, 184 - 215; D. L. Gresh, E. A. Marouf, G. L. Tyler, P. A. Rosen, and R. A. Simpson (1989), Icarus, 78, 131-168; Doyle, L. R., L. Dones, and J. N. Cuzzi (1989), 80, 104-135

79. Showalter, M. R., J. B. Pollack, M. E. Ockert, L. R. Doyle, and B. Dalton (1992) Icarus, 100, 394-411

80. Esposito, L. W. and J. E. Colwell (1989), Nature, 339, 605-607; J. E. Colwell and L. W. Esposito (1990), Icarus, 86, 530-560; also (1990), Geophys. Res. Lett, 17, 1741-1744

81. Weidenschilling, S. J., C. R. Chapman, D. R. Davis, and R. Greenberg (1984), in "Planetary Rings", Univ. of Arizona Press.

82.. Longaretti, P.-Y. (1989) Icarus, 81, 51-73

83. Goldreich, P. and S. Tremaine (1982), Ann. Revs. Astron. Astrophys., 20, 249-283

84. Lissauer, J. J., P. Goldreich, and S. Tremaine (1985), Icarus, 64, 425-434

85. Marley, M. R., and C. C. Porco (1993) Icarus 106, 508-524

86. Humes, D. H. (1980), J. Geophys. Res. 85, 5841-5852

87. Connerney, J. E. P. and J. H. Waite (1984) Nature, 312, 136-138

88. Northrop, T. E. and J. E. P. Connerney (1987) Icarus, 70, 124-137

89. Ip, W.-H. (1983) Icarus, 60, 547-552; Lissauer, J. J. (1984) Icarus, 57, 63-71; Cuzzi, J. N. and R. H. Durisen (1990), Icarus, 84, 467-501

90. Esposito, L. W., A. Brahic, J. A. Burns, and E. A. Marouf (1990); In "Uranus", J. T. Bergstralh, ed. University of Arizona Press

91. Goertz, C. K. and G. E. Morfill (1983) Icarus 53, 219-229

92. Tagger, M., R. N. Henriksen, J. F. Sygnet, and R. Pellat (1990) Astrophys. J. 353, 654-657; also M. Tagger, R. N. Henriksen, and R. Pellat (1990) Icarus, 91, 297-314

93. Goertz, C. K., G. E. Morfill,W. Ip, E. Grün, and O. Havnes (1986), Nature, 320, 141-143

94. Harris, A. W. (1984) in "Planetary Rings", University of Arizona Press

95. Pollack, J. B. (1985) in "Protostars and Planets II", University of Arizona Press; Korycansky, D., P. Bodenheimer, and J. B. Pollack (1991), Icarus, 92, 234-251

96. Nishida, S. (1983) Progr. Theor. Physics, 70, 93-105; Lissauer, J. J. and D. M. Kary (1991), Icarus 94, 126-159

97. Smith, B. A. *et al.* (1981), Science, 212, 163-191; B. A. Smith *et al.* (1982), Science, 215, 504-537; B. A. Smith *et al.* (1986), Science, 233, 43-64; B. A. Smith *et al.* (1989), Science, 246, 1417-1421

98. Lissauer, J. J., S. Squyres, and W. R. Hartmann (1988) J. Geophys. Res. B, 93, 13776-13804; Colwell, J. E., and L. W. Esposito (1992) J. Geophys. Res. E (Planets), 97, 10227-10241; Colwell, J. E., and L. W. Esposito (1993) J. Geophys. Res. E (Planets), 98, 7387-7401

99. Farinella, P., A. Milani, A. M. Nobili, P. Paolicchi, and V. Zappala (1983) Icarus, 54, 353-360

100. Duncan, M., T. Quinn, and S. Tremaine (1988) Astron. J., 328, L69-L73

101. Moses, J. J. (1992) Icarus, 99, 368-383; Lyons, J. R. (1994) B.A.A.S. 26, 1093; also GRL, submitted

102. Dones, L. (1991) Icarus, 92, 194-203

103. Rosen, P. A., G. L. Tyler, E. A. Marouf, and J. J. Lissauer (1991) Icarus, 93, 25-44; Nicholson, P. D., D. P. Hamilton, K. Matthews, and C. Yoder (1992) Icarus, 100, 464-484; Yoder, C. F., S. P. Synnott, and H. Salo (1989) Astron. J. 98, 1875-1889

THERMAL HISTORY OF PLANETARY
MATERIALS IN THE SOLAR NEBULA

T.V. RUZMAIKINA

Lunar and Planetary Laboratory

The University of Arizona, Tucson, AZ 85721

1. Introduction

The current bulk composition of planets has been cast in some extent during the early history of the solar system, associated with the formation and evolution of the solar nebula. Particularly, this stage regulated the abundance and composition of volatiles absorbed by planetesimals and incorporated into planets, and the composition and texture of the primitive meteorites – chondrites.

The history of the solar nebula is usually conditionally divided into three stages: the infall stage, associated with collapse of the presolar cloud and formation (and viscous evolution) of the primordial solar nebula; the further stage of viscous evolution of the solar nebula and planetesimal accumulation; and, finally, the dissipation of the gaseous part of the solar nebula in the region of terrestrial and outer planets and the gas accretion by the cores of the giant planets.

The infall provided the most intense heating of a significant portion of the matter and, in spite of the "shortness" ($\sim 10^5$ yrs), played an important role in the thermal history of the solar system. The maximal temperature of dispersed solids determined composition of volatiles preserved in them, a possibility of survival of the presolar grains, and influenced the oxidation state of both solids and gas.

One of the parameters of which the evolution of the solar nebula depends the most critically is an angular momentum of the presolar cloud (a fraction of the molecular

Earth, Moon, and Planets **67**: 209–215, 1995.

cloud participated in the collapse and joined the Sun or the solar nebula). The larger angular momentum, the larger region over which the gas, falling onto a razor thin solar nebula, is distributed; the radius of this region is called the centrifugal radius. (If the solar nebula has a nonzero thickness and radius larger than the centrifugal radius, then some fraction of the infalling matter will hit the edge of the disk and may join the disk at the distances larger than the centrifugal radius.)

The initial angular momentum of the presolar cloud has an uncertainty of one order or two orders of magnitude, which implies a necessity to investigate the whole range of possible initial angular momenta.

Direct 2D or 3D numerical simulations are restricted at present by angular momenta $(J/M \geq 10^{20}$ cm^2 s$^{-1})$. (Partially these restrictions are imposed by the spatial resolution of numerical codes; for slowly rotating clouds centrifugal forces in the infalling envelope are important only in a relatively close and not well resolved vicinity of the star). According to Boss (1993), the dynamical compression of the solar nebula in such models can result in the temperature ≥ 1500 K in the middle plane of the solar nebula at distances up to $2 - 3$ AU. Further work is required to construct the evolutionary models of the solar nebula and investigate the radial transport from the hot region and its mixing with less heated material.

The lower limit for the angular momentum of the presolar cloud was estimated by the reconstruction of the solar nebula from the present-day structure of the solar system. It is equal to a few times 10^{51} g cm^2 s^{-1} to 10^{52} g cm^2 s^{-1}, or the specific angular momentum $J/M \simeq 10^{18}$ to 10^{19} cm^2 s^{-1} (Weidenschilling 1977, Weissman 1991)).

This paper gives a summary of a study of the thermal history of low-mass solar nebula forming in a collapsing cloud with the initial angular momentum $J = 2 \times 10^{52}$ g cm^2 s^{-1} $(J/M \simeq 10^{19}$ cm^2 s$^{-1})$. (Such a value of the angular momentum might be typical for dense cores of molecular clouds, those rotation was braked before the collapse has started: see, e.g., Safronov and Ruzmaikina 1985). The solar nebula evolution during infall is considered by using a simplified numerical model developed by Ruzmaikina and Maeva 1986, and Ruzmaikina et al 1993, or summarizes the results of these papers. Temperature of presolar silicate dust particles and dust aggregates in the accretional shock (i.e., by the shock produced by the infalling gas at the surface of the solar nebula) were calculated by Ruzmaikina and Ip (1994). The heating of the presolar ices in the shock at the outer solar nebula is a new result, taking into account earlier neglected important effect of the gas cooling in·the postshock region.

2. Model of the Solar Nebula Formation

Let us consider the presolar cloud with the mass 1.1 $M_\odot$ and the angular momentum 2×10^{52} g cm^2/s. We investigate the infall stage of the solar nebula formation, i.e., a stage when embryo protosun and solar nebula have been forming already in the center of the collapsing cloud, but they are still surrounded by the infalling envelope of gas and dust. The solar nebula is considered as a viscous disk with the sources of mass and angular momentum associated with the infall; and with effective viscosity $\nu = 1 \cdot 10^{15} \sqrt{\frac{M}{M_\odot} \frac{R}{1\text{AU}}}$ cm^2/s, which corresponds to conversion of a few percent of the kinetic energy of the infalling gas into the turbulence in the solar nebula. The (magnetic) coupling between the solar nebula and protosun is taken into account by imposing nonzero torque at the inner boundary; the shock, produced by the infalling gas, determines the outer edge of the solar nebula. Calculations start at the moment when a low-mass star-like core, containing a few percent of the total mass, and tiny embryo disk have been formed at the center of the presolar nebula, and the bulk of mass remained in the envelope.

The equations, describing the disk evolution could be reduced to the continuity equation with the source term

$$\frac{\partial \sigma}{\partial t} + \frac{1}{R} \frac{\partial (\sigma R u_R)}{\partial R} = f_a, \tag{1}$$

where u_R is the radial velocity of in the disk

$$u_R = -\left[\frac{3}{j\sigma} \frac{\partial}{\partial R}(j\nu\sigma) + 2R\left(1 - \frac{j_a}{j}\right)\frac{f_a}{\sigma} + \frac{\dot{M}}{M} R \right], \tag{2}$$

f_a is the flux of material from the envelope onto the disk, j_a is the angular momentum of infalling matter, $\sigma(R)$ is the surface density of gas in the disks ($\sigma(R) = \int \rho(R, z)\, dz$, where $\rho(R, z)$ is the spatial gas density), M is the current mass of the forming protosun. These equations are solved numerically, with a second-order accurate explicit scheme. The source functions f_a and j_a are taken from analytical solution of distribution of velocities and density in the infalling envelope (Cassen and Moosman 1981, Ziglina and Ruzmaikina 1991).

The equations governing the vertical structure of the individual rings in the disk are the equation of vertical hydrostatic equilibrium

$$\frac{dP}{dz} = -\frac{GM}{R^3} z \rho, \tag{3}$$

and the equation of the energy balance

$$D_{mech} = D_{rad} + D_{conv}, \tag{4}$$

where P is the gas presure, D_{mech} is the energy released through the shear ($D_{mech} = \frac{9}{4}\frac{GM}{R^3}\nu\rho$ when only the radial shear of the Keplerian motion is taken into account). The radiative transport equation may be written for the frequency-integrated moments, in the approximation of the local thermal equilibrium

$$\frac{dH}{dm} = k_J J - k_B B = D_{mech},\qquad(5)$$

$$\frac{dK}{dm} = k_H H,\qquad(6)$$

where J, H, K are the frequency-integrated moments of the specific intensity, and B is the frequency integrated Plank function, k_J, k_B, and k_H are absorption, mean, Plank mean, and flux mean opacity per unit mass, respectively, $m(z) = \int_z^\infty \rho\, dz$. The equations (3 – 6) for the vertical structure of the disk are solved numerically by an iteration procedure, described by Hubeny (1990).

The shape of the disk was identified with the shape of the shock between the disk and accreting envelope determined by the equation (Elmegreen 1978).

$$\frac{dz}{dR} = \frac{\tan\beta - \frac{R}{z}}{1 + \frac{R}{z}\tan\beta},\qquad(7)$$

where the shock exists, and otherwise with the surface at which density of the infalling gas is equal to the density in the disk; β is an angle between radius vector and normal (to the shock front) component of the velocity ($\beta = \arccos\frac{|v_\perp|}{v} + \arcsin\frac{v_\theta}{v}$), and $v_\perp$ is found from the conservation relation for the impulse, which can be reduced to a balance between dynamical pressure of infalling gas and the thermal pressure in the disk, taking into account fast cooling of the shocked gas resulting in a narrowness of the postshock region. This equation, is integrated from the disk edge inside, checking at every grid point if $v_\perp > c_s$. When the shock disappears we continue by solving of equation $\rho_1 = \rho_2$, (where $\rho_1(z, R)$ is the density of infalling gas and $\rho_2(z, R)$ is the density of the solar nebula at the same height), because the shock disappears and arises again at this surface. (In our approximation $v_\perp = c_s$, at the surface $\rho_1 = \rho_2$). Then we find again that $v_\perp \geq c_s$, specify the position where the shock arises, use it as a new boundary condition, and repeat the procedure again.

3. Results

During infall stage of formation (10^5 yrs), the solar nebula, which started from radius < 1 AU and mass $\approx 10^{-5}$ M$_\odot$ has grown to radius $R_{SN} \geq 50$ AU, while the centrifugal radius for the infalling matter did not exceed $R_{ce} \sim 1$ AU (for $J = 2 \times 10^{52}$ g cm^2/s).

The ratio of the solar nebula mass to the mass of the protosun is < 0.1 during whole stage of infall; the ratio of the maximum height to radius ratio is about 0.1, (see also Ruzmaikina Khatuncev, and Konkina 1993).

Distribution of the radial velocity in the solar nebula depends on efficiency of mixing of newly fallen gas with the solar nebula. In the absence of mixing (when the infalling gas flow around the disk till its centrifugal force balances the gravity, at $R \leq R_{ce}$) the net mass flux in the solar nebula is directed inward in the vicinity of the protosun ($R \leq R_{ce}$), and outward at larger distances. The radial flow in the solar nebula is more complicated if the mixing is efficient (Ruzmaikina and Maeva 1986). It is directed inward in the most part of the solar nebula (except near the edge) at the beginning of the infall, but later an additional region of the outward flow develops at distances between $R_{ce} \leq R \leq 0.5R_{SN}$. In both cases, the outward flow results in spreading of the material from ≤ 1 AU over the asteroid region, and possibly farther. It implies that the radial transport, rather than high temperatures in the solar nebula within several AU, might be a reason for "a hot thermal history" of the matter in the outer part of region of the terrestrial planets. In the most of outer part of the solar nebula the mixing of the falling gas, possessing smaller specific angular momentum than that in the solar nebula, cause inward mass flow. In this region, the strongest heating caused by the shock, produced by the infalling matter at the surface of the solar nebula.

Infalling gas produces strong shocks in two regions - at the vicinity of the star ($R \leq R_{ce}$) and near the edge of the disk. A wide region is situated between them, in which the gas falls under small angles, and the model predicts that the shock should be weak or absent at all. (It could be incorrect if the infalling gas contains inhomogeneities, because denser clumps of the gas penetrate deeper into the solar nebula, and produce strong shock ahead of them at any angle of infall). Note, that the heating of the outer part of the disk by the shock wave makes the IR-spectrum of the disk more flat, and it plausibly could explain an effect observed for some T Tauri stars, a "slow" decrease with the distance from the star of the disk's effective temperature (Beckwith et al 1990).

The shocks can play an important role in the thermal reprocession of dust particles and aggregates, and gas-phase chemical reactions. The peculiarity of the shock in the solar nebula is fast cooling of the gas, (by molecular hydrogen, dipole molecules, and dust emission) resulting in a sharp density gradient in the postshock region, and increase of the gas density up to two orders of magnitude. Subillimeter-size and large dust aggregates, embedded in the infalling gas, have such an inertia that they cross

the region of cooling without deceleration, and are heated by the drag through the
cooled and compressed postshock gas.

Silicate grains, heated to > 1600 K, were melted and then solidified in $\sim 10^3$ s,
plausibly forming chondrules (Ruzmaikina and Ip 1994). The possibility of formation
of such aggregates (chondrule precursors) in the collapsing cloud was first discussed
by Cameron (1978). A recent detailed numerical simulation by Weidenschilling and
Ruzmaikina (1994) reveals a possibility of the formation of fluffy dust aggregates of
appropriate masses in turbulent dense molecular cloud cores, and their survival (or
further growth) during the following collapse.

In the outer part of the solar nebula, where the temperature of the infalling gas
is low enough to preserve presolar organic material and ices, the accretional shock
can cause melting, evaporation, and chemical changes of volatile components of dust
aggregates. Also, the presence of liquid water or vapor could cause oxidation of less
volatile components, such as Fe. The intensity of heating of aggregates and the extent
of region of their evaporation in the outer solar system are also dependent on the size
and structure of the aggregates. In the shock produced by the gas falling normal
to the shock front, millimeter-size dense aggregates could be melted and evaporated
at distances up to 30 AU, and possibly farther. If normalized (for the same rate of
accretion) the extension of the region of evaporation ice aggregates is larger than was
estimated earlier (Mukhin et al 1989, and Lunine et al 1991). This stronger heating
is associated with the fast cooling and compression of the postshock gas – the effect
which was not taken into account in earlier papers. Smaller (say, $\leq 1\mu$m) grains and
very low dense dust aggregates are decelerated faster than the gas cools, and they are
heated to lower temperatures. Such particles preserve interstellar water ice beyond
5 to 10 AU, organics beyond 3 to 5 AU, and silicates beyond 1 AU. The outer solar
system must contain a mixture of interstellar ices, and ices which were evaporated in
the accretional shock and recondensed again.

The general conclusion of this paper is that the presented model can explain the
formation of the low-mass solar nebula with a signature of the high temperature
reprocessing of the matter within inner several AU and the presence of a relatively
unaltered interstellar dust in the outer solar system.

5. References

Beckwith, S.V.W., Sargent, A.I., Chini, R.S., and Güsten, R. (1990) A survey for circumstellar
    disks around young stellar objects, *Astron. J.* **99**, 924 - 945.

Boss, A. P. (1993) Evolution of the solar nebula. II. Thermal structure during nebula formation, *Astrophys. J.* **417**, 351 - 367.

Cameron, A. G. W. (1978) Physics of the primitive solar accretion disk, *Moon and Planets* **18**, 5 - 40.

Cassen, P.M. and Moosman, A. (1981) On the formation of protostellar disks, *Icarus* **48**, 353 - 376.

Elmegreen, B.G. (1978) On the interaction between a strong stellar wind and a surrounding disk nebula, *Moon and Planets* **19**, 261 - 277.

Hubeny, I. (1990) Vertical structure of accretional disks: a simplified analytical model, *Astrophys. J.*, **351**, 632 - 641.

Lunine, J.I., Engel, S., Rizk, B., and Horanyi, M. (1991) Sublimation and reformation of icy grains in the primitive solar nebula, *Icarus* **94**, 333 - 344.

Mukhin, L.M., Ruzmaikina, T.V., and Grechinskii, A.D. (1989) Origin of dust component of Halley's comet, *Kosmicheskie Issledovaniya* **27**, 280 - 285.

Ruzmaikina, T.V. and Maeva, S.V. (1986) Process of formation of the solar nebula, *Astronomicheskii Vestnik* **20**, 212 - 226.

Ruzmaikina, T.V. and Ip, W. (1994) Chondrule formation in radiative shock, *Icarus* **112**.

Ruzmaikina, T.V., Khatuncev, I.V., and Konkina, T.V. (1993) Formation of the low-mass solar nebula, *LPSC XXIV , (Abstracts)* , 1225-1226. .

Weissman, P.R. (1991) The angular momentum of the Oort cloud, *Icarus* **89**, 190 - 193.

Weidenschilling, S.J. and Ruzmaikina, T.V. (1994) Coagulation of grains in static and collapsing protostellar clouds, *Astrophys. J.* **430**, 713 - 726.

Ziglina, I.N. and Ruzmaikina, T.V. (1991) Influx of interstellar material onto the protoplanetary disk, *Astronomicheskii Vestnik* **25**, 53 - 58.

MIGRATION OF BODIES
IN THE ACCUMULATION OF PLANETS

S. I. IPATOV

Institute of Applied Mathematics, Miusskaya Sq. 4, Moscow, Russia

Our investigations of the migration of bodies in the Solar system and the formation of planets were based mainly on the results of computer simulation of the evolution of disks that originally consisted of hundreds of gravitating bodies moving around the Sun (Ipatov 1987, 1993a). The mutual gravitational influence of bodies was taken into account by the Tisserand spheres method, that is two two-bodies problems were considered. Ipatov (1988a) also made analytical studies of the dependence of the time of disk evolution as a function of the number of bodies constituting the disk. For choosing the pairs of encounting (up to the radius of the Tisserand sphere) bodies, Ipatov (1993b) used the probability and deterministic methods. For the probability method, the pairs of encounting bodies were chosen proportionally to the probabilities of their encounters. For the deterministic method, the time to an isolated (from other bodies) encounter of the pair of encounting bodies was minimum. To our opinion, the deterministic method is more physical. Orbits and masses of formed planets, and times to collisions of separate bodies with planets are almost identical for both methods. However, the time of the evolution of disks, consisting of a large number of bodies, and so the time of the formation of the main part of planets' masses obtained by using the deterministic method are ten times less than those obtained by using the probability one. In particular, the time to form 80 % of mass of the Earth did not exceed 10 Myr.

We investigated the evolution of disks that originally consisted of several ring zones corresponding to the feeding zones of the terrestrial planets. Ipatov (1993a) showed that each of these planets incorporated planetesimals from all these zones. After the mass of the Earth's embryo has exceeded 10 % of the present mass of the Earth, the average orbital eccentricities of planetesimals in the feeding zones of the terrestrial planets on the whole exceeded 0.2. Some of these planetesimals penetrated the asteroid belt. Most of the planetesimals that fell onto the Earth underwent collisional evolution. The embryonic masses of unformed terrestrial planets may have exceeded 0.1 of the mass of the Earth.

Ipatov (1987, 1993a) investigated the evolution of various disks corresponding to the feeding zones of giant planets. Most of these disks included almost-formed Jupiter and Saturn. We obtained that average eccentricities of orbits of planetesimals in the feeding zones of Uranus and Neptune exceeded 0.3 during the larger part of disk evolution. A large number of planetesimals

Earth, Moon, and Planets **67**: 217–219, 1995.

from these zones migrated to Jupiter, which ejected them into hyperbolic orbits. The total mass of bodies ejected from the zones of giant planets into hyperbolic orbits may have been ten times as large as the mass of bodies that entered into the planets. The results of our investigations (Ipatov 1987) agree with the values of initial mass of the protoplanet cloud equal to 0.04-0.1 of the mass of the Sun. Jupiter (its nucleus and envelope) might include more ices and rocks than any other planet. The total mass of the bodies from the zone of giant planets, entering the zone of the asteroid belt, could reach tens of Earth's masses. A large amount of water could be delivered to Earth during the accumulation of Uranus and Neptune. The embryos of unformed planets with masses equal to several Earth's masses in the zone of Jupiter and Saturn would be necessary to explain present eccentricities as well as present periods of axial rotation and inclinations of axes of rotation of these planets. Under the influence of migrating bodies the semimajor axis of the orbit of Jupiter may have been shortened by 0.5 AU and those of the other giant planets may have been increased significantly. The embryos of Uranus and Neptune with initial masses equal to several Earth's masses may have originated near the orbit of Saturn and then may have migrated to the present distances from the Sun moving in nearly circular orbits. Some smaller objects may have migrated from the zones of Jupiter and Saturn to the zones of Uranus and Neptune in the same way. The total mass of bodies that penetrated beyond the orbit of Neptune may have reached tens of Earth's masses.

Ipatov (1995b) investigated migration of bodies to the Earth from various regions of the Solar System. We obtained that perihelia or aphelia of orbits of bodies that collided the Earth mainly were near the orbit of the Earth. A large number of objects should exist which orbits lie inside the orbits of Earth and Venus. Most of asteroids of the Amor group should have come from the asteroid belt.

The orbital eccentricity of a body increases significantly only if it encounters several bodies during its evolution. The results of numerical integration of the equations of motion for the plane three-body problem (the Sun and two bodies) showed (Ipatov 1993a) that for initially circular heliocentric orbits and the chaotic variations in orbital elements at $\mu_1 \leq 10^{-5}$ the maximum eccentricity e_{max} usually doesn't exceed $7 - 8 \times \mu_1^{1/3}$, where μ_1 is a ratio of the larger body mass to the Sun mass. For regular orbital variations, e_{max} is smaller. The values of e_{max} and regions of $\varphi_\circ$ and $\varepsilon_\circ = (a_2^\circ - a_1^\circ)/a_1^\circ$ corresponding to various types of orbital variations were investigated by Ipatov (1994) for $10^{-9} \leq \mu_1 \leq 10^{-3}$, where $\varphi_\circ$ is an initial angle with the apice in the Sun between directions to bodies, a_1° and a_2°, are initial values of semimajor axes. These investigations were made for the following types of orbital variations: the motion around triangular points of libration in tadpole and horseshoe synodical orbits, the case of close encounters of bodies, and the chaotic variations in orbital elements when close encounters can't take place. For initially eccentrical orbits and a large time interval, the values of e_{max} depend mostly on initial eccentricities, semimajor axes, and orbital orientations. Ipatov (1981) showed that in the case of close encounters, using the spheres method and choosing the radius of the sphere in an appropriate way, we can obtain almost the same values of e_{max} as those obtained by numerical integration. The "used spheres" method gives even better results, if we consider a larger number of bodies. For the case of three identical bodies circling the Sun, the maximum eccentricities can be several tens of times larger than those for two such bodies (Ipatov 1988b, 1995a).

The results presented above are generally consistent with those obtained by Wetherill (1985), Gladman (1993), and many other scientists. The comparisons of these results were made in our articles listed below.

This work was supported by the Russian Fund of Fundamental Investigations under Grant 93-02-17035.

References

Gladman, B.: 1993, "Dynamics of systems of two close planets." *Icarus*, **106**, N 1, 247-263.

Ipatov, S.I.: 1981, "Computer modeling of the evolution of the plane rings of gravitating particles moving around the Sun." *Sov. Astron.*, **25(58)**, N 5, 617-623.

Ipatov, S.I.: 1987, "Accumulation and migration of the bodies from the zones of giant planets." *Earth, Moon, and Planets*, **39**, N 2, 101-128.

Ipatov, S.I.: 1988a, "Evolution times for disks of planetesimals." *Sov. Astron.*, **32(65)**, N 5, 560-566.

Ipatov, S.I.: 1988b, "Computer simulation of the possible evolution of the orbits of Pluto and bodies of the trans-Neptune belt." *Kinematics Phys. Celest. Bodies*, **4**, N 6, 76-82.

Ipatov, S.I.: 1993a, "Migration of bodies in the accretion of planets." *Solar System Research*, **27**, N 1, 65-79.

Ipatov, S.I.: 1993b, "Methods of choosing the pairs of contacting bodies for investigations of the evolution of discrete systems with binary interactions." *Mathematical Modeling*, **5**, N 1, 35-59 (in Russian).

Ipatov, S.I.: 1994, "Gravitational interaction of two planetesimals moving in close orbits." *Solar System Research*, **28**, N 6 (p. 10-33 in Russian edition).

Ipatov, S.I.: 1995a, "Gravitational interaction of objects moving in crossing orbits." *Solar System Research*, **29**, N 1 (p. 1-13in Russian edition).

Ipatov, S.I.: 1995b, "Migration of small bodies to the Earth." *Solar System Research*, **29**, N 4 (in press).

Wetherill, G.W.: 1985, "Occurrence of giant impacts during the growth of the terrestrial planets." *Science*, **228**, 877-879.

INTERNATIONAL CONFERENCE ON COMPARATIVE PLANETOLOGY

LISTING OF PARTICIPANTS

Name		Affiliation
Michael	A'Hearn	University of Maryland-Dept of Astronomy
Mark	Allen	California Institute of Technology-Planetary Sciences
David	Atkinson	University of Idaho-Dept of Electrical Engineering
Sushil	Atreya	University of Michigan-Dept of Atmospheric, Oceanic & Space Sciences
Kevin	Baines	Jet Propulsion Laboratory
Akiva	Bar-Num	Tel-Aviv University-Dept of Geophysics
Ed	Barker	NASA Headquarters-Code SL
Alexander	Basilevsky	Vernadsky Institute
Robert	Bell	University of California, Los Angeles
Jay	Bergstralh	NASA Headquarters-Code SLC
David	Black	Lunar & Planetary Institute
Stephen	Bougher	University of Arizona-Space Science Bldg.
J. E.	Brandenburg	Research Support Instruments
Michael	Brown	University of California, Berkeley-Dept of Astronomy
Bryan	Butler	California Institute of Technology
N. V.	Bystrova	SAO
James	Campbell	NASA Headquarters-Code SL
Robert	Carlson	Jet Propulsion Laboratory
Moustafa	Chahine	Jet Propulsion Laboratory
Carlson	Chambliss	Kulztown University
Athena	Coustenis	DESPA
Jeff	Cuzzi	Ames Research Center
William	Dent	University of Massachusetts-Astronomy Dept
Charles	Elachi	Jet Propulsion Laboratory
Therese	Encrenaz	DESPA
Kevin	Grazier	University of California, Los Angeles
Ronald	Greeley	Arizona State University-Geology Dept
Samuel	Gulkis	Jet Propulsion Laboratory
Mark	Gurwell	California Institute of Technology
Vicky	Hamilton	Arizona State University-Dept of Geology
John	Hinrichs	University of Hawaii-Manoa
Lon	Hood	University of Arizona
William	Hubbard	University of Arizona
Walter	Huebner	NASA Headquarters
David	Hufnagel	Johnson County Community College

Earth, Moon, and Planets **67**: 221–223, 1995.

INTERNATIONAL CONFERENCE ON COMPARATIVE PLANETOLOGY

LISTING OF PARTICIPANTS

Name		*Affiliation*
Andrew	Ingersoll	California Institute of Technology
James	Kaler	University of Illinois
Jeff	Kargel	US Geological Survey
Bill	Kaula	University of California, Los Angeles-Dept of Earth & Space Sciences
Ken	Klaasen	Jet Propulsion Laboratory
Eugene	Levy	University of Arizona-Faculty of Science
Lynn	Lewis	Ball Aerospace
Jonathan	Lunine	University of Arizona-Space Science
Jim	Lyons	California Institute of Technology
Teemu	Makinen	Finnish Meteorological Institute
Mikhail	Marov	North Carolina State University-Mars Mission Research Center
B. H.	Mauk	Johns Hopkins University-Applied Physics Laboratory
Fridman	Maximovich	Institute of Astronomy
Albert	Metzger	Jet Propulsion Laboratory
Michael	Mickelson	Denison University-Dept of Physics & Astronomy
Diedrich	Mohlmann	DLR-Institut fur Ranmsimulation
Hari	Nair	California Institute of Technology-Geological & Planetary Sciences
Jim	Nations	Jet Propulsion Laboratory
William	Nellis	Lawrence Livermore National Laboratory
William	Newman	University of California, Los Angeles
Neil	Nickle	Jet Propulsion Laboratory
Sandra	Orellana	California State University, Dominguez Hills
Fritz	Osell	Leeward Community College-Astronomy Dept
Tobias	Owen	University of Hawaii
Timothy	Pham	Jet Propulsion Laboratory
Jurgen	Rahe	Jet Propulsion Laboratory
Nick	Renzetti	NASA Headquarters-Code SLD
Elizabeth	Roettger	Adler Planetarium
J.	Russell	Brown University-Dept of Geological Sciences
Tamara	Ruzmaikina	University of Arizona-Lunar & Planetary Laboratory
Gerald	Schubert	University of California, Los Angeles-Institute of Geophysics & Dept of Earth & Space Sciences
Michael	Schulz	Lockheed Palo Alto Research Lab-Space Sciences Dept
Tero	Shli	Ames Research Center
Richard	Simpson	Stanford University

INTERNATIONAL CONFERENCE ON COMPARATIVE PLANETOLOGY

LISTING OF PARTICIPANTS

Name		Affiliation
Martin	Slade	Jet Propulsion Laboratory
Paul	Spielman	Ridgecrest, Ca.
Thomas	Spilker	Jet Propulsion Laboratory
Walter	Spjeldvik	NASA Headquarters-Code SSM
Paul	Spudis	Lunar & Planetary Institute
David	Stevenson	California Institute of Technology-Geological & Planetary Sciences
Edward	Stone	Jet Propulsion Laboratory
Nobuya	Tajima	University of Tokyo-Dept of Earth & Planetary
Richard	Thiessen	Washington State University
Dimitris	Tsintikidis	University of Iowa-Dept of Physics & Astronomy
Gunter	Wuchterl	Universitat Wien
Lawrence	Zanetti	Johns Hopkins University

LISTING OF PARTICIPANTS

THE KLUWER LATEX STYLE FILE

Kluwer Academic Publishers has developed a special style file for authors who want to submit LATEX articles. KLUWER.STY is a general LATEX style file which is used for all Kluwer journals, irrespective of the publication's size or layout. (The specific journal characteristics are added later during the production process.) Authors are kindly requested always to use KLUWER.STY when creating a LATEX article for a Kluwer journal.

Instruction File
Although KLUWER.STY is very similar to the ARTICLE.STY and uses many of the standard LATEX commands, there are some differences. These are explained in the accompanying instruction file - KAPINS[number].TEX

Getting the Kluwer Style File

KLUWER.STY is offered at a number of servers around the world. Unfortunately, those are unauthorized copies and authors are strongly advised not to use them. Kluwer can only guarantee the integrity of files obtained directly from Kluwer.

Gopher-server, E-mail or Air Mail
Authors can obtain KLUWER.STY and the instruction file from Kluwer Academic Publishers' GOPHER-server. This free service is available at:

GOPHER.WKAP.NL
IP-Number: 192.87.90.1
WWW URL: gopher://gopher.wkap.nl/

The stylefile may be requested from:

Kluwer Academic Publishers, Editorial Department,
P.O.Box 17, 3300 AA Dordrecht, The Netherlands.
Telephone: (0)78-392392, Fax: (0)78-392254
E-mail: EDITDEPT@WKAP.NL

The files can be sent either by e-mail or on diskette. Don't forget to mention the journal's name, your e-mail number, and postal address.

Earth, Moon and Planets **67**: 225–226, 1995.

Submitting Manuscripts

Please send your completed LaTeX article on diskette, together with the appropriate number of hard copies to the address listed in the *Instructions to Authors* of your journal.

Via E-mail
Experience has shown that sending articles via e-mail can sometimes result in lacunas appearing in the article. That's why Kluwer prefers to receive LaTeX articles on diskettes, accompanied by the hard copies. If you must send the LaTeX article via e-mail, don't forget to send the requisite number of hard copies by air mail.

Questions

Should you have any questions or encounter problems using KLUWER.STY, please contact Kluwer Academic Publishers for assistance.

AUTHOR INDEX

(Volume 67)

Anderson, B. J., 175

Atreya, S. K., 71

Basilevsky, A. T., 47

Bougher, S. W., 31

Brandenburg, J. E., 35

Coustenis, A., 95

Cuzzi, J. N., 179

Edgington, S. G., 71

Encrenaz, T., 77

Gautier, D., 71

Greeley, R., 13

Hood, L. L., 131

Hunten, D. M., 31

Ipatov, S. I., 217

Kargel, J. S., 101

Kaula, W. M., 1

Levy, E. H., 143

Möhlmann, D., 115

Nakagawa, Y., 67

Owen, T. C., 71

Potemra, T. A., 175

Roble, R. G., 31

Ruzmaikina, T. V., 209

Schulz, M., 161

Spilker, T. R., 89

Tajima, N., 67

Wuchterl, G., 51

Zanetti, L. J., 175

CONTENTS TO VOLUME 67

Vol. 67   Nos. 1–3   1994/1995

COMPARATIVE PLANETOLOGY WITH AN EARTH PERSPECTIVE

Edited by M. T. CHAHINE, M. F. A.'HEARN and J. RAHE

Foreword	vii
Introduction	ix–x
WILLIAM M. KAULA / Formation of the Terrestrial Planets	1–11
RONALD GREELEY / Geology of Terrestrial Planets with Dynamic Atmospheres	13–29
S. W. BOUGHER, D. M. HUNTEN and R. G. ROBLE / CO_2 Cooling in Terrestrial Planet Thermospheres	31–33
JOHN E. BRANDENBURG / Constraints on the Martian Cratering Rate Based on the SNC Meteorites and Implications for Mars Climatic History	35–45
A. T. BASILEVSKY / Factors Controlling Volcanism and Tectonism in Solar System Solid Bodies	47–49
G. WUCHTERL / Giant Planet Formation: A Comparative View of Gas-Accretion	51–65
NOBUYA TAJIMA and YOSHITSUGU NAKAGAWA / Giant Planet Formation: Dynamical Stability of a Massive Envelope	67–69
S. K. ATREYA, S. G. EDGINGTON, D. GAUTIER and T. C. OWEN / Origin of the Major Planet Atmospheres: Clues from Trace Species	71–75
T. ENCRENAZ / The Chemical Atmospheric Composition of the Giant Planets	77–87
THOMAS R. SPILKER / NH_3, H_2S, and the Radio Brightness Temperature Spectra of the Giant Planets	89–94
A. COUSTENIS / Titan's Atmosphere and Surface: Parallels and Differences with the Primitive Earth	95–100
J. S. KARGEL / Cryovolcanism on the Icy Satellites	101–113
D. MÖHLMANN / Formation of Satellite and Ring Systems: Comparative Aspects	115–129
L. L. HOOD / Frozen Fields	131–142
E. H. LEVY / Planetary Dynamos	143–160
MICHAEL SCHULZ / Planetary Magnetospheres	161–173

L. J. ZANETTI, T. A. POTEMRA, and B. J. ANDERSON / Boundary
  Determinations from Low Frequency Magnetic Field Measurements   175–178

JEFFREY N. CUZZI / Evolution of Planetary Ringmoon Systems   179–208

T. V. RUZMAIKINA / Thermal History of Planetary Materials in the Solar
  Nebula   209–215

S. I. IPATOV / Migration of Bodies in the Accumulation of Planets   217–219

List of Participants   221–223

The 'Kluwer' LaTeX Style File: Instructions for Authors   225–226

Author Index   227

Volume Contents   229–230